Praise for *Not Even Wrong*

"Peter Woit presents an authoritative, sobering, and very readable history of a scientific and sociological phenomenon that is largely unprecedented in the history of science. From a physics perspective, what is known as 'string theory' remains primarily the 'hope' for a theory. Nevertheless it has dominated theoretical particle physics as well as the popular consciousness as few other notions have, all the while without ever making a single falsifiable prediction about nature! Readers interested in a more balanced understanding of modern theoretical physics are sure to benefit."

—LAWRENCE KRAUSS, author of *The Physics of Star Trek*, and
*Hiding in the Mirror: The Mysterious Allure of Extra
Dimensions, from Plato to String Theory and Beyond*

"*Not Even Wrong* is an authoritative and well-reasoned account of string theory's extremely fashionable status among today's theoretical physicists. I regard it as an important book."

—ROGER PENROSE, author of *The Road to Reality*

"[A]rgue[s] passionately that string theory has played itself out. In the two and a half decades since it first captivated physicists, despite thousands of published papers and the expenditure of billions of dollars, there is no proof whatsoever that string theory is correct. Not one prediction of the theory has been experimentally testable."

—*Boston Globe*

"Woit offers some intriguing ruminations on the relationship between physics and mathematics. . . ."   —*New York Times* Book Review

"[Woit] explores the interface between maths and physics, concluding that mathematicians view string theory as physics and physicists regard it as mathematics. The two communities are at odds over whether string theory is a series of abstract puzzles or whether it says something about the real world. The idea that beauty can point to scientific truth served Einstein well. Sadly for science, it may have misled a later generation of theoretical physicists."　　　*—The Economist*

"[L]ively and entertaining."　　　*—Discover* magazine

"The story of how a backwater of theoretical physics became not just the rage but the establishment has all the booms and busts of an Old West mining town."　　　*—Scientific American*

"[A] tightly argued, beautifully written account. . . ."

—Publishers Weekly

"[A] call to arms for physicists to pursue multiple paths in search of truth, not funding."　　　*—New Scientist*

"That string theory abandoned testable prediction may be its ultimate betrayal of science."　　　*—Wall Street Journal*

NOT EVEN
WRONG

THE FAILURE OF STRING THEORY AND
THE SEARCH FOR UNITY IN PHYSICAL LAW

PETER WOIT

BASIC
BOOKS

A Member of the Perseus Books Group
New York

Hardcover published in 2006 by Basic Books
A Member of the Perseus Books Group
Paperback published in 2007 by Basic Books

Books published by Basic Books are available at special discounts for bulk purchases
in the United States by corporations, institutions, and other organizations. For more
information, please contact the Special Markets Department at the Perseus Books
Group, 2300 Chestnut Street, Suite 200, Philadelphia, PA 19103, or call (800) 255-1514,
or e-mail special.markets@perseusbooks.com.

Library of Congress Cataloging-in-Publication Data
Woit, Peter.
   Not even wrong : the failure of string theory and the search for unity in physical law /
Peter Woit.
     p. cm.
   Includes bibliographical references and index.
   ISBN-13: 978-0-465-09275-8 (alk. paper)
   ISBN-10: 0-465-09275-6 (alk. paper)
   1. String models. 2. Supersymmetry. 3. Quantum theory. 4. Superstring theories.
5. Physical laws. I. Title.

   QC794.6.S85W65 2006
   539.7'258--dc22

                          2006013933

Paperback ISBN-13: 978-0-465-09276-5
Paperback ISBN-10: 0-465-09276-4

08 09 / 10 9 8 7 6 5 4 3 2

FOR ELLEN

CONTENTS

PREFACE

Throughout the year 2005, physicists all over the world celebrated the centennial of Albert Einstein's revolutionary discoveries of 1905, which included the special theory of relativity and the quantum nature of light. These discoveries led quickly to dramatically new and powerful physical theories, and continuous progress was made throughout much of the twentieth century on using them to understand the fundamental nature of matter and physical forces. By 1973, an extremely successful theory of elementary particles and their interactions was in place, a theory that soon became known as the "standard model."

After 1973, life grew much more difficult for particle physicists as evidence mounted that their field was in danger of becoming a victim of its own success. The standard model left many questions about fundamental physics still open, but new generations of experiments produced results that agreed with the model precisely, giving no hints about where to look for something better. Particle physics entered a period quite unlike any in its earlier history, one that continues to this day. In a few years, results will become available from a new, higher-energy particle accelerator now under construction near Geneva. There is some reason to hope that these will finally give some indication of how to get beyond the standard model, but this is far from a sure thing.

Throughout the late seventies and early eighties, particle theorists explored a wide range of new ideas for how to better understand and extend the standard model. One of these was rather radical, since it involved replacing the whole notion of pointlike elementary particles by one-dimensional objects that acquired the name "strings." In this "string theory," the strings were supposed to be so small that they appeared to be points as far as all feasible experiments were concerned. In order for strings to look at all like known particles, a more complicated version of string theory called "superstring theory" was required.

The main motivation behind superstring theory was that it held out hope of being able to address one of the questions the standard model left unanswered, that of how to deal with the gravitational force. Einstein's general relativity provided a simple and beautiful geometrical theory of gravity, but technical problems arise when one tries to combine this with the quantum-theoretic principles on which the standard model is based. Initially, superstring theory garnered little attention, and only a very small number of physicists worked on it. This changed dramatically after the summer of 1984, a moment that superstring theorists now refer to as the "First Superstring Theory Revolution."

At that time, a technical calculation showing that certain potential problems canceled in very specific cases attracted the attention of Edward Witten, the leading figure in particle theory. He began intensively working on superstrings, enthusiastically promoting the idea and helping it receive wide attention. Within a year or so, a large number of theorists joined him, and superstring theory quickly became the dominant topic of research in the field.

Particle theory has a long history of being successfully pursued in a somewhat faddish manner, partly due to the influence of unexpected experimental results. Certain new ideas get a lot of attention, leading in a short period either to significant progress, or, more commonly, to abandonment as the community moves on to the next thing. Superstring theory broke this pattern dramatically, leading the field into a period with no real historical parallel. From the beginning, there was both no experimental evidence for superstrings and various obvious problems in the way of ever being able to use them to make experimental predictions. The theory required postulating the existence of many extra unobserved dimensions, and by different choices of the properties of these extra dimensions, one could get just about anything one wanted.

Remarkably, the lack of any progress in achieving a predictive version of the theory that could be tested by experiment did not lead to theorists giving up on the superstring idea. Instead, it achieved a sort of critical mass, as a whole new research field grew up, largely disconnected from the rest of physics. The great complexity and poorly understood nature of superstring theory provided many topics for theorists to work on, while at the same time avoiding any possibility of showing that the idea was wrong. A large number of popular science articles, books, and even TV programs promoted superstrings to the general public. By 2004, there was an undergraduate textbook devoted to the subject.

While the failure of superstring theory as a unified theory of elementary particles grew ever more difficult to ignore, up until the turn of the century, consistent progress was being made toward better understanding some of the implications of the superstring idea. This sort of internal progress slowed down dramatically, and in recent years attempts to connect up superstrings with reality have taken a rather bizarre turn. Many string theorists have become convinced that superstring theory inherently must allow an astronomically large number of physical possibilities, so many that it is difficult to see how the theory can ever be tested. Normally, this sort of conclusion would cause physicists to abandon a theory, but some theorists have instead chosen to claim this as a virtue. They see the existence of this "landscape" of possibilities as justifying the use of something called the "anthropic principle." Maybe we really live in a "multiverse" of different possible universes, and the one we are in has the particular laws of physics we observe just because those laws are among the few possibilities hospitable to life. This way of thinking about physics does not seem to lead to any falsifiable predictions, and so is one that physicists have traditionally considered to be unscientific.

The willingness of some physicists to give up on what most scientists consider the essence of the scientific method has led to a bitter controversy that has split the superstring theory community. Some superstring theorists continue to hold out hope that a better understanding of the theory will make the landscape problem go away. Others argue that physicists have no choice but to give up on long-held dreams of having a predictive theory, and continue to investigate the landscape, hoping to find something about it that can be used to test the idea experimentally. The one thing both camps have in common is a steadfast refusal to acknowledge the lesson that conventional science says one should draw in this kind of circumstance: if one's theory can't predict anything, it is just wrong and one should try something else.

The physicist Wolfgang Pauli was known for his often less than polite criticism of the work of some of his colleagues. He would sometimes exclaim "wrong" (*falsch*) or "completely wrong" (*ganz falsch*) when he disagreed with someone. Near the end of his life, when asked his opinion of an article by a young physicist, he sadly said "it is not even wrong" (*Das is nicht einmal falsch*).[1] The phrase "not even wrong" is a popular one among physicists, and carries two different connotations, both of which Pauli likely had in mind. A theory can be "not even wrong" because it is so incomplete and ill-defined that it can't be used to make firm predictions whose failure would show it to be wrong. This has been the situation of superstring theory from its beginnings to the present day.

This sort of "not even wrong" is not necessarily a bad thing. Most new theoretical ideas begin in this state, and it can take quite a bit of work before their implications are well enough understood for researchers to be able to tell whether the idea is right or wrong. But there is a second connotation of "not even wrong": something worse than a wrong idea, and in this form the phrase often gets used as a generic term of abuse. In the case of superstring theory, the way some physicists are abandoning fundamental scientific principles rather than admit that a theory is wrong is something of this kind: worse than being wrong is to refuse to admit it when one is wrong.

One topic of this book is the controversial situation of superstring theory described above, but it also tells a much more positive story. While the last twenty years have been trying ones for high-energy particle physics, they have been very good ones for the interaction of mathematics and physics. The mathematics used in the standard model is incredibly rich, and this has led to many fruitful new questions and ideas that have had a huge influence on mathematics.

While string theory may turn out to have been a disaster for physics, it has led to much wonderful new mathematics. This is an intricate story, involving the most advanced areas of mathematics and physics, and thus not easy to make accessible to a wide audience. But it is these new and difficult ideas that may ultimately be the important developments of the past thirty years, rather than the superstrings, extra dimensions, and multiple universes that have received so much popular attention.

Throughout the history of physics, there have been only a few examples of successful theoretical developments driven by concerns of mathematical beauty and coherence rather than by experimental data. One of the most famous of these is Einstein's general relativity, his theory of gravity based on sophisticated mathematical ideas about geometry. It is much easier to make progress when hints from experiment are available to tell one which direction to take, but the lack of such hints just means that one must take a more difficult road. In recent years, theoretical physics has contributed a great deal to mathematical research; perhaps in the future mathematics will return the favor.

ABOUT THIS BOOK

Much of the story I am telling is uncontroversial, and most experts on the subject would more or less agree with how it is being told here. On the other hand, the reader should be aware that later parts of this book are about topics that are quite controversial, and my point of view on these topics is by no means a majority one. Readers will have to judge for themselves how much credence to give to my arguments, and this is one reason for including here both some unusually technical material as well as a fair amount of detail about the background and experiences of the author.

The more technical chapters have been written without the use of equations, and an attempt has been made as much as possible both to avoid technical vocabulary and to offer at least some sort of explanation of vocabulary that can't be avoided. These choices lead to a certain lack of precision that experts may find trying. While the hope is that many nonexperts will be able to follow much of these chapters, the large number of difficult and abstract concepts involved is likely to make this quite a challenge in some places.

Such chapters have been structured to begin with an introductory section summarizing in general terms what is at issue and how it fits into the story of the book. Professional physicists and mathematicians are quite used to the idea that one cannot hope always to follow the entirety of a technical discussion, and that one therefore needs to be ready to skip ahead to where things again get less demanding. Just about all readers should find this tactic necessary at one point or another. For those who want to understand some of the more technical chapters in greater detail, a section at the end of these chapters gives an annotated list for suggested further reading. A real understanding of many of the topics discussed can't be achieved by reading a few pages of text, but requires traveling a rather difficult path. I hope at least to describe the landmarks on that path and point readers to where such a journey really starts should they choose to embark on it.

Much of this book is about history, and an accurate description of this history, were it possible, would require a very different and much larger volume. What appears here is more of a quick sketch, ignoring the fine details of exactly who discovered what, when. In place of this there is often just a short description of the physicists or mathematicians whose names have conventionally been attached to various discoveries. This by no means should be taken to indicate that these are necessarily the actual discoverers. In a course on particle physics I took at Harvard from the Spanish physicist Alvaro De Rujula, whenever he introduced a concept with someone's name attached to it, he would generally say something like the following: "This is the so-called Weinberg angle, which of course was discovered not by Weinberg, but by Glashow." On one occasion, after introducing a named concept he stopped for a while and seemed to be thinking deeply. Finally he announced that, as far as he knew, strangely enough, this concept actually seemed to have been discovered by the person whose name was attached to it.

This book deals with both the history and current state of particle physics from a slightly unusual point of view, and to explain this

I should mention some history of a personal nature. My earliest memories of being concerned with the issues to be discussed here go back to the first years of the 1970s, to hours spent poring over every book about astronomy I could find in the local public library. At some point I came across the subject of astrophysics, in particular that part of the subject that studies the structure of stars by writing down and then solving equations for the temperature, pressure, and composition of the interior of a star. That one could hope to understand in such a detailed and precise way exactly what was going on in the unimaginable interior of a star fascinated me, but was also mystifying. The equations in the books I was reading were expressed in a mathematical language I could not understand, and were derived from physical laws I knew nothing about. I began trying to study the necessary mathematics and physics to make sense of those equations.

As I learned some basic ideas about calculus and elementary physics, one of the first striking lessons was that mathematics and physics were intertwined in a very complex way. Mechanics, the part of elementary physics that deals with the motions of particles and the forces that cause these motions, is based on Newton's laws, which require calculus for their expression. Newton developed calculus and mechanics at the same time, and the two subjects are so completely entangled that one cannot properly understand one without understanding the other. Using the language of calculus, Newton's laws are exceedingly simple and clear statements about the way that at least part of the world works.

As I took more physics books out of the library, I began to find out about areas of physics other than mechanics, and I soon came across and fell in love with something that has fascinated me to this day: quantum mechanics. While the equations of Newton's classical mechanics refer to easily visualizable quantities such as the position and velocity of particles, the fundamental equation of quantum mechanics, Schrödinger's equation, concerns a mathematical entity completely out of the realm of ordinary experience: the wave function. While the wave function and Schrödinger's equation for it seem to have no relation to anything one can visualize, they have allowed physicists to understand and predict precisely an incredible variety of physical phenomena that take place on the distance scale of the size of an individual atom.

One book that made a strong impression on me was Werner Heisenberg's memoir *Across the Frontiers*,[2] in which he tells the story of his experiences during the 1920s, the early days of quantum mechanics. He describes long debates with his friends about the nature of physical reality held during hikes in the local mountains. The basic ideas at issue were those that soon led him, Erwin Schrödinger, and others to the explosion of new ideas about physics that was the birth of quantum mechanics in 1925. Later on, after I had learned more about events in Germany between the wars, the image of Heisenberg and others in his youth group marching around the mountains to attend large inspirational gatherings began to take on more troubling aspects. Part of the appeal of quantum mechanics to me was its peculiar character of being a kind of esoteric practice. Through long study and deep thought, one could hope to arrive at an understanding of the hidden nature of the universe. Unlike other popular exotic religious or mind-altering activities of the time, this sort of search for enlightenment appeared to be both much more solid and something for which I actually had some talent.

When I went off to college at Harvard in 1975, I soon found that the physics department there was in a state of great excitement, in some ways similar to that which had characterized physics soon after the birth of quantum mechanics precisely fifty years earlier. The standard model had recently been formulated, and experimental evidence for it was beginning to pour in. This theory was a quantum field theory, a more sophisticated version of the quantum mechanics I was just starting to study seriously. My undergraduate advisor was Sheldon Glashow, and in the office two doors down was Steven Weinberg, with whom he would later share a Nobel Prize for their independent work on part of the standard model. One of the young postdocs was David Politzer, a codiscoverer of the other main piece of the theory. He would soon be joined by another postdoc, Edward Witten, from Princeton, who was destined to be the next leader of the field. Great things had happened and more were expected imminently from this impressive array of talent.

During my college years I spent a formative summer working on a particle physics experiment at the Stanford Linear Accelerator Center, and a lot of time trying to figure out what quantum field theory was all about. I graduated in 1979 with a hazy idea of the subject and some basic notions about the standard model, and went on directly to doctoral study at Princeton. The physics department faculty there included David Gross, who with his student Frank Wilczek had played a crucial role in the development of the standard model. It was soon to include Witten, who returned to Princeton as a tenured professor directly from his postdoc, skipping over the usual tenure track. For me, this was a time of getting down to learning quantum field theory seriously, and beginning to try to do some original work. For the field as a whole, it was the beginning of a frustrating period. Many ideas were floating around about how to go beyond the standard model, but none of them seemed to be working out successfully.

I left Princeton in 1984 to spend three years as a postdoctoral research associate at the Institute for Theoretical Physics at SUNY Stony Brook. My arrival there coincided with a period that came to be known as the "First Superstring Revolution," a series of events that will be described later in this book, and which marked a great change in the field of particle theory. By the last of my three years at Stony Brook, it became clear to me that someone interested in mathematics and quantum field theory wouldn't have much of an immediate future in a physics department unless he or she wanted to work on the new superstring theory. This impression was confirmed by the negative results of a job search for a second postdoc.

Since my research interests involved the parts of quantum field theory closest to mathematics and I did not want to do superstring theory, it seemed that it would be a good idea to try my luck looking for employment among the mathematicians. I moved back to Cambridge, where the physics department at Harvard let me use a desk as an unpaid visitor, and the mathematics department at Tufts hired me as an adjunct to teach calculus. From there I went on to a one-year postdoctoral research associate position at the Mathematical Sciences Research Institute at Berkeley, followed by a four-year non-tenure-track junior faculty appointment in the mathematics department at Columbia.

This change of fields from physics to mathematics turned out to be a wise move, and I have now been at Columbia in the math department for more than sixteen years. Currently, I am happily in the nontenured but permanent faculty position of lecturer, with one of my main responsibilities being making sure the department's computer system keeps functioning properly. I also teach classes at the undergraduate and graduate levels, as well as continue to do research in the area of the mathematics of quantum field theory.

My academic career path has been rather unusual, and I am very much aware that it has been based on a significant amount of good luck. This began with the good fortune of having parents who could afford to send me to Harvard. It continued with being in the right place at the right time to take advantage of an uncommon opportunity to work in an excellent math department surrounded by talented and supportive colleagues.

The experience of moving from physics to mathematics was somewhat reminiscent of a move in my childhood from the United States to France. Mathematics and physics each have their own distinct and incompatible languages. They often end up discussing the same thing in mutually incomprehensible terms. The differences between the two fields are deeper than simply that of language, involving very distinct histories, cultures, traditions, and modes of thought. Just as in my childhood, I found that there is a lot to learn when one makes such a move, but the extra effort is compensated by an interesting bicultural point of view. I hope to be able to explain some of what I have learned about the complex, continually evolving relationship between the subjects of physics and mathematics and their corresponding academic cultures.

When I sat down to write about some of these topics, I began by trying to write out a short history of quantum mechanics and particle theory. My perspective was different from that of most exercises of this kind, which typically ignore the role of mathematics in this story. As I looked more deeply into some of the standard books on the subject, I noticed something intriguing: one of the major figures in the small circle of those who discovered and developed quantum theory was actually a mathematician, Hermann Weyl. During the very short period during which physicists were working out quantum mechanics in 1925 and 1926, Weyl was in constant communication with them, but was himself in a burst of inspiration doing the purely mathematical work that was to be the high point of his career. The field of mathematics Weyl was involved with at the time is known as group representation theory, and he was well aware that it was the right tool for understanding part of the new quantum mechanics. Physicists were almost entirely baffled by Weyl's mathematics and how it fitted into the new quantum theory, even after Weyl quickly wrote a book containing alternate chapters on quantum theory and representation theory.[3] For many years the book was considered a classic, but most physicists probably read just half of the chapters.

Group representation theory is the mathematical expression of the notion of a "symmetry," and understanding of the importance of this notion slowly grew among particle theorists throughout the 1950s and 1960s. By the 1970s, courses on group representation theory involving parts of Weyl's work had become a standard part of the theoretical physics curriculum. From then on, particle theory and mathematics have interacted closely in a very complex way. Explaining the twists and turns of this story is one of the main goals of this book.

The positive argument of this book will be that historically, one of the main sources of progress in particle theory has been the discovery of new symmetry groups of nature, together with new representations of these groups. The failure of the superstring theory program can be traced to its lack of any fundamental new symmetry principle. Without unexpected experimental data, new theoretical advances are likely to come about only if theorists turn their attention away from this failed program and toward the difficult task of better understanding the symmetries of the natural world.

1

Particle Physics at the Turn of the Millennium

At the end of his closing talk at a conference in Kyoto in 2003, the theoretical physicist David Gross finished with a dramatic flourish, quoting from a speech given by Winston Churchill. In Gross's version, near the end of his life Churchill rose to give a campaign speech, in which he exhorted his listeners, "Never, never, never, never, never give up." This story is similar to one repeated by many people, but the source of the quotation is a speech Churchill gave at Harrow School during the war, which contains the following words: "This is the lesson: never give in, never give in, never, never, never, never—in nothing, great or small, large or petty—never give in except to convictions of honour and good sense."

The conference in Kyoto was called "Strings 2003," and it brought together several hundred theoretical physicists who work on string theory, a set of ideas that has dominated theoretical particle physics for the last two decades. Gross is one of the world's most prominent theorists; after a very distinguished career at Harvard and Princeton, he is now director of the Kavli Institute for Theoretical Physics, at Santa Barbara. He was to share the 2004 Nobel Prize in physics for work done in 1973 that was of huge significance for the field of particle physics. What disturbed Gross so much that he would invoke the elder statesman Churchill and the words he used to rally his country during the dark days of World War II?

His concern was that recent developments in string theory may be leading many physicists to abandon the traditional central goal of theoretical physics: to understand the physical world in terms of a simple compelling theory, and use this to make predictions that test this understanding. Gross quoted from a section of Einstein's autobiographical writings, written late in his life, at the age of 67:

> I would like to state a theorem which at present can not be based upon anything more than upon a faith in the simplicity, i.e., intelligibility, of nature . . . nature is so constituted that it is possible logically to lay down such strongly determined laws that within these laws only rationally completely determined constants occur (not constants, therefore, whose numerical value could be changed without destroying the theory).[1]

Einstein is stating the creed that Gross and most theoretical physicists believe: there is a single set of simple underlying laws that describe how the universe works, and these laws are uniquely determined. There are no extra parameters that determine the theory; once one gets the right idea about what the laws are, there are no additional numbers that one needs to specify in order to write them down. Gross's Nobel Prize was awarded for his 1973 codiscovery of an extremely successful theory of one of the forces experienced by certain kinds of elementary particles, and this theory has exactly the uniqueness property that Einstein believed in. This theory has no free parameters that can be adjusted to fit experiments, and yet it accurately predicts a wide range of different experimental results.

This abandonment of Einstein's creed that so worried Gross has taken the form of an announcement by several leading theorists that string theory is compatible with an unimaginably large number of different possible descriptions of the world, and as a result, perhaps the only predictions it can make are those that follow from the "anthropic principle." The anthropic principle is essentially the idea that our very existence puts constraints on what physical laws are possible. These must be such that intelligent beings such as ourselves could somehow evolve. If a huge number of different universes exist, all with different physical laws, we are guaranteed to be in one of the ones where intelligent life is possible.

One of the leading proponents of this point of view is Leonard Susskind, a professor at Stanford and one of the codiscoverers of string theory, who explains,

> Mostly physicists have hated the idea of the anthropic principle; they all hoped that the constants of nature could be derived from the

beautiful symmetry of some mathematical theory . . . Physicists always wanted to believe that the answer was unique. Somehow there was something very special about the answer, but the myth of uniqueness is one that I think is a fool's errand . . . If there were some fundamental equation which, when you solved it, said that the world is exactly the way we see it, then it would be the same everywhere. On the other hand you could have a theory which permitted many different environments, and a theory which permitted many different environments would be one in which you would expect that it would vary from place to place. What we've discovered in the last several years is that string theory has an incredible diversity—a tremendous number of solutions—and allows different kinds of environments. A lot of the practitioners of this kind of mathematical theory have been in a state of denial about it. They didn't want to recognize it. They want to believe the universe is an elegant universe—and it's not so elegant. It's different over here, it's that over here. It's a Rube Goldberg machine over here. And this has created a sort of sense of denial about the facts about the theory. The theory is going to win, and physicists who are trying to deny what is going on are going to lose.[2]

Susskind's vision of the universe as a complicated, inelegant Rube Goldberg machine that is the way it is because otherwise we wouldn't be here has gained an increasing number of adherents, and he has written a popular book on the subject entitled *The Cosmic Landscape: String Theory and the Illusion of Intelligent Design*.[3] Gross refers to the anthropic point of view as a "virus"[4] that has infected many physicists, who show no signs of ever recovering from the disease. He tells the story of his younger colleague at Santa Barbara Joe Polchinski, who at one point felt that anthropic reasoning was so nefarious he would resign his professorship rather than engage in it, but now has gone over to the other side. Two years after Strings 2003, in a public talk at Strings 2005, in Toronto, Susskind was describing the ongoing controversy as a "war" between two groups of physicists, also comparing it to a "high-school cafeteria food fight." He claimed that his side was winning, with Gross's in retreat, and accused his opponents of being in "psychological denial" and engaged in "faith-based science." At a panel discussion held during the

Toronto conference, the panel of leaders in the field split evenly over the anthropic issue, while the audience voted 4 or 5 to 1 against Susskind's point of view.

How did particle physics get itself into its current state where some of its most prominent practitioners question whether their colleagues have given up on science? Have they? Why has there been so little progress in this subject for the last quarter-century, and where should one look for ways to change this situation? The following chapters will describe some of the history that has led particle physics to its current predicament. Since 1973, the field has failed to make significant progress, and in many ways has been the victim of its own success. The reasons for this failure will be examined, and an attempt will also be made to extract lessons from the history of previous successes that may indicate a more promising way forward.

The Instruments of Production

The bourgeoisie cannot exist without constantly
revolutionizing the instruments of production . . .
—KARL MARX, THE COMMUNIST MANIFESTO

The central concern of this book is the recent history and present
state of theoretical particle physics, especially in its relationship to
mathematics. But to understand anything about this, one has first to
understand the material conditions that are fundamental to particle
physics research. Particle accelerators and detectors are the "instruments of production" that are used to create the base of experimental
data upon which all theorizing about elementary particles is built.
The continuing improvement and refinement of these experimental
tools is what has driven progress in particle theory during much of
the past century. This chapter will explain the basic principles governing how accelerators work, describe some of their history and present state, and finally consider what the prospects are for their future.

BASIC PRINCIPLES

Before it is possible to explain any of the basic physical principles
needed to understand how experimental particle physics is done, certain fundamental conventions of how to describe measurements
have to be set. This is the question of what system of measurement
units to use. There are many different units in use in different subfields of physics, but particle physicists have one preferred set of
units, sometimes referred to as "God-given" or "natural" units.
These units are chosen to take advantage of basic features of special
relativity and quantum mechanics, getting rid as much as possible of

constants that depend on the choice of measurement units by choosing units in such a way that these constants can be set equal to one.

A fundamental postulate of special relativity is that space and time are linked together so that the speed of light is always constant, no matter which reference frame it is measured in. This is what makes the subject paradoxical from the point of view of everyday experience: if I try to move at high speed in the same direction as a beam of light, no matter how fast I go, the light will always be moving away from me at the same speed. The equations of special relativity simplify when units of measurement for space and time are chosen to be such that the speed of light is equal to one. For example, one way of doing this is to note that light travels 300,000 kilometers in a second, so it travels about a foot in a nanosecond (the prefix "nano" means "one billionth"). As a result, measuring lengths in feet and times in nanoseconds would make the speed of light about one. Setting the speed of light equal to one determines the choice of units used to measure time in terms of the choice of units used to measure space, and vice versa.

Perhaps the most famous equation related to Einstein's special relativity is the $E = mc^2$ equation relating energy (E), mass (m), and the speed of light (c). Note that using units in which the speed of light is set equal to one simplifies this to $E = m$, so energy and mass become equal in the context described by this equation. As a result, particle physicists use the same units to measure energy and mass.

While special relativity links together the way spatial dimensions and the time dimension are measured, quantum mechanics links together energy and time measurements. This will be explained in greater detail later on, but two basic facts about quantum mechanics are that:

1. There is a mathematical entity called a "state vector" that describes the state of the universe at a given time.
2. Besides the state vector, the other fundamental mathematical entity of the theory is called the Hamiltonian. This is an operator on state vectors, meaning that it transforms a given state vector into a new one. Operating on a general state vec-

tor at a given time, it tells one how the state vector will change during an infinitesimal additional time period. In addition, if the state vector corresponds to a state of the universe with a well-defined energy, the Hamiltonian tells one what this energy is.

The fact that the Hamiltonian simultaneously describes the energy of a state vector as well as how fast the state vector is changing with time implies that the units in which one measures energy and the units in which one measures time are linked together. If one changes one's unit of time from seconds to half-seconds, the rate of change of the state vector will double and so will the energy. The constant that relates time units and energy units is called Planck's constant (after the physicist Max Planck) and conventionally denoted by the letter h. It is generally agreed that Planck made an unfortunate choice of how to define the new constant he needed, since it almost always enters equations divided by a factor of two times the mathematical constant π ($= 3.14159265\ldots$). As a result, physicists prefer to work with Planck's constant h divided by two times π, a constant conventionally written as an h with a bar through it, $\hbar$, and called h-bar. Particle physicists choose their units so as to make $\hbar$ equal to one, and this fixes the units of time in terms of the units of energy, or vice versa.

With these choices of the speed of light and $\hbar$, distance units are related to time units, and time units are related to energy units, which in turn, as described before, are related to mass units. The standard convention of particle physics is to express everything in energy units, and thus one just has to pick a single measurement unit, one that determines how energies are expressed. Here, theorists bow to the experimentalists, who long ago found it most convenient to measure energies in electron volts. An electron volt (abbreviated eV) is the energy an electron picks up as it moves between two metal plates that have a voltage difference of one volt between them. Once one has chosen to measure energies and masses in units of eV, then the choice of constants described earlier means that time and space (which are measured in inverse units to energy) are measured in "inverse electron volts" or $(eV)^{-1}$.

Table 2.1

Energy	Example
0.04 eV	Energy of atoms in air at room temperature
1.8–3.1 eV	Energy of photons of visible light
100–100 000 eV	X-rays
20 keV	Kinetic energy of electrons in a television monitor
More than 100 keV	Gamma rays
511 keV	Mass of an electron
1–10 MeV	Energies produced in nuclear decays
105 MeV	Mass of a muon
938 MeV	Mass of a proton
93 GeV	Mass of a Z boson
1 TeV	Energy in each proton in a beam at the Tevatron

To provide a feel for what these energy units are like, Table 2.1 gives the values of various masses and energies corresponding to several different particle physics phenomena (some to be described in more detail later on), all in electron volts. The standard abbreviations for a large number of electron volts include 10^3 eV = 1 keV (kilo electron volt), 10^6 eV = 1 MeV (mega electron volt), 10^9 eV = 1 GeV (giga electron volt), and 10^{12} eV = 1 TeV (tera electron volt).

All the energies in this table are those of a single particle or photon, so on everyday scales they are very small amounts of energy, with 1 TeV being about the same as the kinetic energy (energy of motion) of a slow-moving ant. There is a much larger energy that theorists sometimes consider, the "Planck energy" of about 10^{19} GeV. This is conjecturally the energy scale at which quantum effects of gravity become important. It is a much more significant amount of energy, corresponding roughly to the chemical energy in an automobile's tank of gasoline.

In the units we are discussing, the unit of distance is the inverse electron volt, which in more conventional units would be about a micron (10^{-6} meters, a millionth of a meter). Time is also measured in inverse electron volts, and this unit of time is extremely short, roughly 4×10^{-15} seconds. Since energies are measured in eV and distance in $(eV)^{-1}$, particle physicists tend to think of distances and energies interchangeably, with one being the inverse of the other. The energy corresponding to the mass of a proton is 1 GeV, a billion electron volts. Since this energy is a billion times larger than an electron volt, the corresponding distance will be one billion times smaller or $10^{-9} \times 10^{-6} = 10^{-15}$ meters. One can think of this distance as being the characteristic size of the proton.

Particle physicists equivalently refer to their investigations as involving either very short distance scales or very high energy scales. Typical physical processes under study involve something that happens at some particular approximate distance or approximate energy, and this is said to be the distance or energy "scale" under study. In accelerators the total energy of the particles one is colliding together sets the energy scale one can study. Investigating shorter and shorter distances requires higher and higher energies, and at any given time the fundamental limit on the experimental information one can gather about elementary particles comes from the technological limits on the energies of particles in one's experiments.

EXPERIMENTAL PARTICLE PHYSICS: A QUICK HISTORY

The history of experimental particle physics is by now a quite long and complex one; this section will give a quick sketch of some of this history. The fundamental experimental technique of particle physics is to bring two particles close together and then watch what happens. The simplest way to do this is to begin by producing a beam of energetic particles in one way or another, and then accelerating the particles to high energy in some sort of accelerator. The beam of high-energy particles is then aimed at a fixed target, and one uses a detector of some sort to see what particles come out of the region where the beam hits the target.

A simple example of this concept is behind the design of a television set. In a cathode-ray tube television design, a beam of electrons (the cathode rays) is accelerated by high voltages toward a target, which is the back of the screen of the TV. Magnetic fields are used to control the beam, in which the electrons reach energies of about 20,000 electron volts. When the beam hits the screen, collisions of the electrons with atoms in the screen produce reactions that lead to the emission of photons of light, which are then detected by the eyes of the TV viewer watching the front of the screen. So the TV is an accelerator with an electron beam, and the detector that analyzes the results of the collisions with the target (the screen) is the human eye.

The collisions going on in the TV screen cause changes in the energy levels of the atoms in the screen, and as a result, a TV might be useful for studying the physics of atoms. If one is interested in even smaller scales or higher energies, a TV is of no use, since the electron beam does not have enough energy to disrupt the atom sufficiently to get at more fundamental physics. To see what happens when electrons collide not with the atom as a whole, but with its constituents (the nucleus and the electrons bound to the nucleus), much higher energies than those in a cathode ray tube are needed.

During the past century, many different possible sources of more-energetic particles were investigated. The first of these sources was naturally occurring radioactivity, for example the radioactive decay of radium, which produces alpha particles (helium nuclei) with an energy of about 4 MeV, or 200 times that of cathode-ray tube electron beams. In 1910, Ernest Rutherford, working at Manchester, England, was the first to discover that most of the mass of an atom is contained in a very small nucleus. He did this by sending a beam of alpha particles produced by radium through a thin sheet of mica. The alpha particles were deflected off the atoms in the mica in a scattering pattern. He could measure this pattern by having an experimenter observe the flashes caused by the alpha particles as they hit a screen coated with zinc sulfide. This pattern indicated that the alpha particles were colliding with something very small, something much smaller than an atom.

Rutherford thus had at his disposal a 4-MeV beam of alpha particles and, as detector, the zinc sulfide screen, which flashed when hit. The next technological advance also occurred in 1910, with the development of the cloud chamber by Charles Wilson. This much more sophisticated particle detector works by quickly reducing the pressure inside a glass chamber, at which point water vapor condenses along the track of ionized particles left by an energetic particle traveling through the chamber. Being able to see the tracks of all charged particles involved in a collision provides a great deal more information about what has happened than that provided by the flashes seen in Rutherford's experiment.

The years after 1910 saw the discovery of a different source of energetic particles, "cosmic rays" coming from the skies. These were particles with energies mostly in the range of a few hundred MeV, but sometimes extending much higher. Experimental particle physics up until the late 1940s was dominated by the task of sorting out the nature of the cosmic rays. Experiments involved observing the particle collisions created by incoming cosmic rays hitting either the atmosphere or a fixed experimental target. Ultimately, it turned out that most cosmic rays are caused by energetic protons hitting the upper atmosphere, creating a shower of pions, muons, and electrons that make up most of what experimenters can observe at ground level. Improvements in these cosmic ray experiments were driven by the construction of better and better detectors, including the Geiger counter and photographic emulsions. These detectors were taken to mountaintops or sent up in balloons to get as many of the most-energetic collisions as possible. This period saw the discovery of many new elementary particles, including the positron in 1932, the muon in 1937, and charged pions and kaons in 1947.

Cosmic rays provide a rather weak and uncontrolled beam of particles with energies of hundreds of MeV or higher, with the beam becoming much weaker at higher energies. Particle physicists very much wanted to gain access to much more intense high-energy particle beams whose energy and direction could be precisely controlled. To achieve this required finding new techniques for accelerating large numbers of particles to high energy. The first such particle accelerator was designed and built by John Cockcroft and Ernest Walton at the Cavendish Laboratory in Cambridge in 1930. This machine used a 200-kilovolt transformer, and was able to accelerate a beam of protons to 200 keV. By 1932, they had reconfigured their accelerator to send the beam through a sequence of accelerating stages, reaching a final energy of 800 keV. The year 1931 saw the appearance of two additional accelerator designs that could reach similar energies. One worked by building up an electrostatic charge, and was designed by Robert Van de Graaff; the other design, by Rolf Wideroe, used a radio-frequency alternating voltage.

The alternating voltage design was adapted by Ernest Lawrence and collaborators at Berkeley, who constructed the first "cyclotron" in 1931. In a cyclotron, the particle beam is bent by a magnetic field and travels in a circle, being accelerated by an alternating voltage each time it goes around the circle. Lawrence's first cyclotron reached an energy of 80 keV, and by mid-1931 he had built one that could produce a beam energy of over 1 MeV. This machine had a diameter of only eleven inches, but over the next few years Lawrence was able to scale up the design dramatically. By late 1932, he had a 27-inch cyclotron producing a 4.8-MeV beam, and in 1939 a 60-inch one with a 19-MeV beam. These machines were becoming increasingly expensive, since they required larger and larger magnets to bend the higher and higher energy beams into a circle. Lawrence needed good fundraising as well as scientific skills. By 1940 he had a promise of $1.4 million from the Rockefeller Foundation to finance a 184-inch diameter machine that could reach 100 MeV. But the war intervened, and this machine's magnet ended up being diverted to be used by the Manhattan Project for uranium enrichment needed to make the Hiroshima bomb.

After the war and the success of the Manhattan Project, physicists were at the peak of their prestige, and they reaped the benefits of a dramatic increase in funding of their projects. Lawrence quickly took advantage of the situation, realizing that "There [is] no limit on what we can do but we should be discreet about it."[1] His laboratory at Berkeley had operated before the war on an annual budget of $85,000, but immediately after the war he was able to increase that to $3 million, partly through the help of the Manhattan Project's director, General Leslie Groves.

At higher energies, the effects of special relativity required changing the design of the cyclotron to that of a "synchrocyclotron," in which the frequency of the accelerating voltage changed as the particles were accelerated. By November 1946, Lawrence had the large Manhattan Project magnet back in civilian use in a 184-inch diameter machine producing a beam with an energy of 195 MeV. Although the cosmic ray physicists were still ahead at actually discovering particles, this situation was soon to change. After the discovery of charged pions in cosmic rays in 1947, the neutral pion was discovered at Lawrence's lab in 1949.

Higher-energy accelerators could not be built using a single magnet, but instead used a doughnut-like ring of smaller magnets. This design was called a "synchrotron," and in 1947, the Atomic Energy Commission approved the construction of two of them. One was called the Cosmotron, and was built at Brookhaven National Laboratory, on Long Island. It achieved an energy of 3 GeV in 1952. The second was built at Berkeley, and called the Bevatron. It had a proton beam of 6.2 GeV, and was completed in November 1954. The late 1950s was the heyday of accelerator construction, with large machines in operation or under construction at more than a dozen locations around the world. The Russians built a 10-GeV proton synchrotron at Dubna in 1957, providing yet another perceived challenge to the technological supremacy of the United States in the same year as the launch of the first Russian Sputnik. From then on, funding for high-energy physics in the United States was to increase dramatically for the next few years, before leveling off in the mid-1960s.

After the war, several European nations joined together to form a joint organization to perform nuclear research. This European Organization for Nuclear Research (Centre Européen de Recherche Nucléaire), known as CERN, was founded in 1952, and soon began building a laboratory near Geneva. The first large accelerator at CERN, the PS (for Proton Synchrotron) was completed in 1959 and operated at an energy of 26 GeV. Very soon thereafter a similar machine was put into operation at Brookhaven, the AGS (for Alternating Gradient Synchrotron), which could reach 33 GeV. The 1960s saw ever-larger machines being constructed, although now in smaller numbers due to their huge cost. In 1967, the Soviet Union built a 70 GeV machine at Serpukhov, and in the same year a new laboratory, Fermilab (named to honor the Italian physicist Enrico Fermi), was founded in Batavia, Illinois, about 45 miles west of Chicago. Construction was begun there on a new accelerator that would be two kilometers in diameter. The Fermilab accelerator was completed in 1972, and had an energy of 200 GeV, increased to 500 GeV by 1976. Meanwhile, at CERN the SPS (Super Proton Synchrotron), capable of reaching 400 GeV, was finished in 1976.

The dominant detector during the late 1950s and early 1960s was the bubble chamber, which was first perfected in the mid-1950s. It was basically a vessel containing liquid hydrogen under pressure. When the pressure was quickly reduced, the liquid became superheated, and trails of bubbles would form along the paths of charged particles. Large bubble chambers such as the 72-inch diameter one at Berkeley and the 80-inch one at Brookhaven were quite expensive to build and to operate. The photographs of tracks that they produced required much laborious human effort to analyze, although in later years the process was partially automated.

The machines mentioned so far were all proton accelerators. There was a parallel development of electron synchrotrons, which included a 1.2-GeV one at Caltech (1956), a 6-GeV one at Harvard (1962), and a 10-GeV one at Cornell (1968). Electron accelerators were less popular than the proton ones, since they had to be run at lower energy, and unlike protons, electrons are not strongly interacting particles, so they could not directly be used to study the "strong interaction." The reason electron synchrotrons run at lower energies is that when the paths of high-energy electrons are bent into a circle by magnets, the electrons give off large amounts of X-ray synchrotron radiation. As a result, energy must be continually pumped back into the electron beam. To get around this problem, a new laboratory was built in Menlo Park, California, near Stanford University, called the Stanford Linear Accelerator Center (SLAC), with its centerpiece a large linear electron accelerator. The SLAC machine is three kilometers long and reached its design energy of 20 GeV in 1967. The accelerator runs in a line from close to the San Andreas fault at the base of the hills near Stanford, eastward toward San Francisco Bay. It has been said that in case of a major earthquake, the laboratory may have to be renamed SPLAC (for Stanford Piecewise-Linear Accelerator Center).

In high-energy accelerators, the beam particles carry a great deal of momentum, and by the law of conservation of momentum, this must be conserved in a collision. As a result, most of the energy of

The Stanford Linear Accelerator Center (SLAC)

the collision goes into the large total momentum that the products of the collision have to carry. The actual energy available to produce new particles grows only as the square root of the beam energy, so the 500-GeV proton accelerator at Fermilab could provide only about 30 GeV for new-particle production. Early on, many physicists realized that if one could collide two accelerator beams head on, the net momentum would be zero, so none of the energy in the beams would be wasted. The problem with this is that the density of particles in a beam is quite low, making collisions of particles in two intersecting beams rather rare.

Accelerators that collide two beams together are now called colliders, although at first they were often referred to as storage rings, since many particles must first be injected and then stored in the accelerator ring before collisions can begin. Several electron–electron and electron–positron colliders were constructed during the 1960s, culminating in one called ADONE at Frascati, Italy, which had 1.5 GeV per beam for a total collision energy of 3 GeV. An electron–positron collider built at SLAC, called the Stanford Positron Electron Asymmetric Rings (SPEAR), had 3-GeV beams and was completed in 1972 (its first physics run was in spring 1973). Later on, this machine was to play a crucial role in the dramatic events of 1974 that will be described later in this book and that came to be known as the "November Revolution." SPEAR was still being used in 1978, and was responsible for providing me with summer employment, working on an experiment called the "Crystal Ball" that was being installed there at the time.

SPEAR was built in a parking lot near the end of the long linear accelerator, which was used to inject particles into the ring. Since it had been impossible to get approval for its construction out of the standard mechanisms for capital funding from the Atomic Energy Commission, it ultimately was built for $5 million out of SLAC's operating funds. This kind of funding required that there be no permanent buildings. As a result, the accelerator ring ran in a tunnel made by placing concrete shielding blocks in the parking lot, and the machine was operated and data analyzed in various nearby trailers.

Particle accelerators and detectors are impressively sophisticated-looking pieces of equipment, and the contrast between this equipment and the ramshackle structures at SPEAR was striking. Many years later I was in an art gallery in SoHo and noticed a show of very large photographs whose subjects were oddly familiar. It turned out that the photographs were of parts of the Crystal Ball experiment. Evidently, its aesthetic aspects had impressed the photographer.

Increasingly large electron–positron colliders were built during the 1970s and 1980s, culminating in the Large Electron Positron (LEP) collider at CERN. This was a huge machine, built in a tunnel 27 km in circumference straddling the French–Swiss border. It began operation in 1989 at a total energy of 91.2 GeV, and operated until November 2000, when it was finally shutdown after having reached a total energy of 209 GeV. At 209 GeV, the particles in LEP lost 2 percent of their energy to synchrotron radiation each time they went around the ring. Running the machine used an amount of electrical power about 40 percent as large as that used by the city of Geneva. Doubling the energy of a ring the size of LEP increases the power needed by a factor of 16, so it seems very likely that no higher-energy electron–positron ring will be built anytime soon, since the cost of the power to run it would be prohibitive.

The first collider to use proton beams was a proton–proton collider called the Intersecting Storage Ring (ISR) built at CERN and commissioned in 1971. It ran until 1983, reaching a total energy of 63 GeV. The next major advance was the revamping of CERN's SPS accelerator in 1981 into a proton–antiproton collider with a total energy of 540 GeV. A collider at Fermilab called the Tevatron became operational in 1983 and began doing physics in 1987 with an energy of 1.8 TeV. This was the first accelerator to use superconducting magnets, which were necessary to achieve the very high magnetic fields required to bend the trajectory of the beam into a circle 6.3 km in circumference.

Detector technology made huge advances during this period as detectors grew into ever larger and more complex instruments using very sophisticated electronics and many layers of different particle-detection technologies. Teams of more than a hundred physicists were involved in the design, construction, and operation of each of these huge arrays, whose price tag could be a sizable fraction of the cost of one of the very large accelerators. This cost limited the number of detectors that could be built, and the social organization of experimental particle physics changed as larger and larger numbers of physicists were working on smaller and smaller numbers of experiments.

While a large ring was built and operated successfully at Fermilab, other new accelerator projects did not fare as well. Ground was broken in 1978 for a 4-km tunnel to be used by an 800-GeV proton–proton collider at Brookhaven called ISABELLE. ISABELLE was a new design using superconducting magnets; technical problems with these magnets slowed its construction. By 1983 the competing collider at CERN was already operational, and the decision was made to abandon the ISABELLE project. The finished tunnel was already in place, but was kept empty for many years until recently it has been put into use to house a machine called the Relativistic Heavy Ion Collider (RHIC) that studies the collisions of heavy nuclei.

After the ISABELLE project was ended, the decision was made to stop work on upgrading the accelerator at Fermilab and devote resources instead to a far more ambitious new plan to construct something to be called the Superconducting Super Collider (SSC). The SSC was to be an 87-km ring and new laboratory complex at a site near Waxahachie, Texas. It was designed to produce a total energy of 40 TeV. This would represent a large jump from the existing highest-energy accelerator, the Tevatron, which ran at 1.8 TeV. The decision to proceed with the project was made in January 1987 at the highest levels of the Reagan administration. After hearing Department of Energy experts make their pitch for SSC funding, Reagan recalled a phrase from his days as a sports reporter, "Throw deep!"[2] and approved the plan. He was then told, "Mr. President, you're going to make a lot of physicists ecstatic," to which he responded, "That's probably fair, because I made two physics teachers in high school very miserable." This decision was ultimately to make a lot more physicists very miserable.

Construction began in 1991 with an original cost estimate of $4.4 billion. As construction proceeded, design changes were made, and the estimated cost of the machine increased to $8.25 billion. The Bush and Clinton administrations continued to support the program, but opposition to it was increasing. This opposition was fed by scientists outside of high-energy physics, who worried that the SSC would crowd out other science spending. Many in Congress were unhappy with such a large chunk of government "pork" going to one district in Texas. By the autumn of 1993, some were estimating that the total cost might climb to $11 billion, and Congress voted to cancel the project. At the time, $2 billion had been spent and 22km of the tunnel had been excavated.

The disastrous impact of the cancellation of the SSC on the particle physics community in the United States is hard to overestimate, and it continues to this day. Novelist Herman Wouk recently published a novel entitled *A Hole in Texas*,[3] whose protagonist is an experimentalist traumatized by the SSC experience. American experimental particle physicists had agreed to "throw deep" with the SSC in a gamble that it would restore their field to a preeminence it had already mostly lost to the Europeans. When this gamble failed, there was no fall-back plan, and the only remaining high-energy U.S. project was to continue to improve the intensity of the beam at the Tevatron. The high-profile political loss in Congress was a definitive victory for those who opposed any large amount of government domestic spending on pure research of the kind represented by particle physics. The ascendancy of these political forces meant that there was no possibility of replacing the SSC project with another one of anywhere near its cost.

CURRENT ACCELERATORS

At the present time, the worldwide number of high-energy accelerators in operation or under construction is extremely small, smaller than at any time since their invention before the Second World War. Those few still being used have not had significant increases in their energy in many years. LEP, the large electron–positron collider at CERN, has been dismantled, and its 27-km tunnel will be used for a new proton–proton collider to be called the Large Hadron Collider (LHC).

The highest-energy machine in operation today is the Tevatron at Fermilab, an accelerator that first went into operation more than eighteen years ago. In 1996, it was shut down for a $300-million five-year overhaul that was designed to increase the number of particle collisions that could be produced. Besides the energy of its beams, the most important characteristic of any collider is its "luminosity." The luminosity of a collider is a measure of how many particles are in the beam and how small the interaction region is in which the beams coming from opposite directions are brought together inside a detector. For any given physical process that one might like to study with a collider, doubling the luminosity makes it happen twice as often. The study of many aspects of particle interactions requires the ability to collect as many rarely occurring events as possible. The complex behavior of high-energy particle beams makes increasing the luminosity as much an art as a science, and typically it can take a year or more from the time beams are first stored in a collider until it is operating reliably at the highest luminosity of which it is capable.

When the Tevatron was turned back on in March 2001 after the upgrade, its energy had been slightly increased (to a total of 1.96 TeV), and the expectation was that its luminosity would soon be about five times higher than before. This turned out to be much more difficult to achieve than anyone had imagined, and it took more than a year just to get the luminosity of the machine back to the level it had reached before being shut down. The hope had been that by the end of 2002, experiments would have observed about 15 times more collisions than during the earlier Tevatron runs, but instead, at that time only about the same number of collisions as before had been seen. By 2005, four years after the recommissioning, the luminosity was six times higher than before the upgrade, but only a quarter of what had been expected to be achieved by that time. Fermilab still hopes to continue improving the luminosity of the Tevatron and ultimately to collect a large number of new collisions, but the experience has been very painful and costly, making all too clear how difficult the technical problems associated with large accelerators actually are.

The Tevatron is an extremely complex and sensitive instrument, with a very large number of things that can go wrong. In November 2002 the beam was unexpectedly lost, and on investigation it turned out that the source of the problem had been an earthquake, 6000 miles away in Alaska. Two detectors are in operation at the Tevatron, named the Collider Detector Facility (CDF) and the DZero (the name refers to the interaction region where it is located). These are huge, sophisticated, and expensive installations, each employing a team of about six hundred physicists. Much of the experimental effort goes into the complex task of analyzing the vast amounts of data produced, a task that can go on for many years after the data has been collected. Fermilab expects to keep the Tevatron operating until around the end of this decade.

Besides the Tevatron, the only other accelerator now operational at anywhere near its energy is the Hadron–Electron Ring Accelerator (HERA) at the Deutsches Elektronen-Synchrotron (DESY) laboratory near Hamburg, Germany. This is a proton–electron collider,

6.3 km in circumference, with a 27.5-GeV electron beam and an 820-GeV proton beam. The net energy available for particle production in collisions is about 300 GeV. HERA began operation in 1992, and has allowed the performance of many kinds of experiments that were not possible with electron–positron or proton–antiproton colliders, but its energy is significantly lower than that of the Tevatron.

Much lower energy electron–positron colliders are in use at Cornell and at SLAC. The machine at SLAC collides 9-GeV electrons with 3.1-GeV positrons and is somewhat of a special-purpose instrument, designed to study matter–antimatter asymmetries in systems involving the bottom quark. Fermilab also has some accelerators with lower energy than the Tevatron, which are used both to inject particles into the Tevatron and to produce beams of particles for other experiments. There are now two separate experiments in operation at Fermilab designed to study neutrinos. The first of these is called MiniBoone, the "Mini" since this is the first stage of a possible large experiment, the "Boone" for Booster Neutrino Experiment, since it uses a 700-MeV neutrino beam produced by an 8-GeV proton synchrotron called the Booster. MiniBoone has been in operation since 2002 on the Fermilab site and is expected to report results within the next year or so.

The second neutrino experiment at Fermilab is known as NUMI/MINOS. NUMI is an acronym for Neutrinos at the Main Injector, MINOS for Main Injector Neutrino Oscillation Search. The Main Injector is a 150-GeV proton synchrotron also used to inject beams into the Tevatron, but used in this case to produce a neutrino beam whose energy can be adjusted in the range from 3 to 15 GeV. The MINOS experiment consists of two detectors. The first, the near detector, is at Fermilab, and the second, the far detector, is 735 km away deep underground in the Soudan mine in Minnesota. By comparing the behavior of neutrino interactions at the two detectors, MINOS will be able to study how neutrinos "oscillate" between different types. The experiment began operation early in 2005 and is scheduled to continue until the end of the decade.

American funding for experimental particle physics has been flat in constant dollars for many years now, even declining during the past year, being kept at the same level with no adjustment for inflation. Each year a total of roughly $775 million is spent on high-energy physics in the United States, with about $725 million of this coming from the Department of Energy and another $50 million from the National Science Foundation. A similar amount is spent each year in Europe, with the bulk of it used to fund CERN and DESY. Much of the CERN budget and some of the U.S. budget in recent years has gone to financing ongoing construction of the LHC (Large Hadron Collider) at CERN, a project toward which the United States is making a $530-million contribution spread over eight years. The Department of Energy budget for FY 2006 includes a 3 percent cut in funding for high-energy physics, and budget prospects for the next few years are uncertain (although a significant increase for FY 2007 has just recently been proposed). The huge U.S. fiscal deficits are putting intense pressures on discretionary spending for scientific research, and high-energy physics has not been high on the federal government's list of scientific priorities.

ACCELERATORS: FUTURE PROSPECTS

The only major new accelerator under construction at the present time—one on which just about all the hopes of particle physicists are riding—is the LHC at CERN. This machine is a proton–proton collider with a total energy of 14 TeV. Unlike the Tevatron, which collides protons and antiprotons, the LHC will collide protons with protons in order to avoid the luminosity problems that come from being unable to produce a sufficiently intense beam of antiprotons.

The LHC will be going into the 27-km tunnel that had been used by LEP. The project was first approved by CERN in 1994, and the latest estimate of when the first beam will be turned on is summer 2007, with first collisions and data being taken by the detectors later that year. If all goes well, the first physics results may begin to arrive during 2008. There will be two main detectors that are part of the project, called Atlas (A Toroidal LHC ApparatuS) and the Compact Muon Solenoid (CMS), each of which employs about two thousand physicists from over thirty different countries. Each of the detectors is expected to produce roughly one thousand terabytes of data per year.

The total cost of the LHC and its detectors is roughly $6 billion, with the detectors costing a billion dollars or so each. Even though part of the cost of the project is being borne by the United States and others, it is so large that the CERN budget has had to be structured in such a way that no funding for any other large projects will be possible through 2010.

While the main constraint on the energy of an electron–positron ring is the problem of synchrotron radiation loss, for a proton ring this is not much of an issue, since protons are so much heavier than electrons. Instead, the problem is the strength of the magnetic field needed to bend the paths of the protons and keep them moving in a circle. The LHC will use 1200 superconducting dipole magnets for this purpose, each with a very high magnetic field strength of 8.4 tesla.

What are the prospects for even higher proton–proton collision energies than those achievable at the LHC? There are really only two viable ways of getting to higher energies: a bigger ring or higher magnetic fields. The energy of a ring scales linearly with its size and the magnetic field, so one could double the energy of the LHC to 28 TeV either by building a ring twice as large or by finding a way to make magnets with twice the field strength. The SSC 40-TeV design required an 87-km ring, the main reason for its high price tag. Superconducting magnets have been built that can achieve field strengths of 16 tesla, but magnets that can actually be used in an accelerator must be conservatively designed, since they need to be highly reliable and capable of being produced at reasonable cost. They also need to be capable of remaining superconducting even in the presence of the large amount of synchrotron radiation coming out of the accelerator. Various designs have been proposed for a VLHC (very large hadron collider), but they all involve very large rings, up to 233 km in size. The immediate political prospects for such a project in the United States are not very promising, given the experience with the SSC. It seems likely that these VLHC designs will not be seriously pursued until after data begin to arrive from the LHC. If the

LHC makes new discoveries that provide a strong physics case for a new machine at a somewhat higher energy, then the argument for a VLHC may be convincing.

An added difficulty in designing any higher-energy collider is that one must increase the luminosity as the energy is increased. At fixed luminosity the rate of "interesting" collisions (those in which a large amount of energy is exchanged) falls off as the square of the energy. So if one doubles the energy of one's accelerator, one needs four times the luminosity to produce the same number of interesting events. As was seen at the Tevatron, higher luminosities are hard to achieve, and even if achieved, lead to new technical problems such as that of radiation damage to parts of detectors near the interaction region.

A new electron–positron ring with energy higher than LEP's 209 GeV is not a realistic possibility due to the synchrotron radiation energy loss problem. Any higher-energy machine would have to be a linear collider consisting of two linear accelerators built end-to-end, one accelerating electrons, the other positrons. Some experience with this kind of design exists, since from 1989 to 1998 the SLAC linear accelerator was operated as the SLAC Linear Collider (SLC), simultaneously accelerating positron and electron beams to about 47 GeV. As they came out of the end of the accelerator, the beams were separated and magnets were used to send them through curved paths that ultimately brought the beams together head-on. In recent years, several designs for possible linear colliders have been pursued, the most advanced of which was a design for a machine to be called TESLA, and constructed at DESY in Hamburg. TESLA is an acronym for TeV Energy Superconducting Linear Accelerator, chosen in homage to the inventor Nikola Tesla, after whom the unit of magnetic field strength is also named.

The TESLA design would require a complex 33 km in length to produce collisions with an energy of 500 GeV (upgradeable to 800–1,000 GeV) and cost very roughly at least $4–5 billion. While there are no synchrotron radiation losses in a linear collider, the beam is not continually recirculated as in a ring, so all the energy that goes into accelerating the beams ends up being lost when they collide. While operating, TESLA would use something like 200 megawatts of power, roughly the same as a city of 200,000 people. The effort to design an electron–positron collider has recently been reorganized as the ILC (International Linear Collider) project, bringing together the previous design efforts of TESLA and various other groups. The physicists involved hope to have a final design completed and ready for a decision to start construction if funding can be found sometime around 2010.

During 2001, the High Energy Physics Advisory Panel for the Department of Energy and the National Science Foundation consulted a large part of the particle physics community and tried to come up with a long-range plan for U.S. high-energy physics. Their report was presented to the federal government in January 2002, and its main recommendation was that the next large high-energy physics project should be the construction of an electron–positron linear collider. The panel considered two main scenarios, one in which the collider would be constructed in the United States, which would require a 30 percent increase in the U.S. particle physics budget, and another in which it would be built outside the United States, which would require a 10 percent increase. The prospects for either a 10 percent or 30 percent increase are not promising, since either would be a big change from the pattern of recent years, which has been a constant or declining budget for high-energy physics research in the United States. If recent trends continue, the U.S. experimental particle physics community will soon face stark choices. After the LHC with its much higher energy becomes operational, by 2010 running the Tevatron collider will no longer make sense, and all experimental work at the highest energies will have moved to Europe. Unless some way is found to fund the ILC and put it at Fermilab, the future of that laboratory will be at issue and American experimental high-energy physics is likely to become something of a backwater, able to carry out only special-purpose lower-energy experiments such as those involving only the study of neutrinos.

CERN is pursuing a different sort of technology for a linear accelerator, known as CLIC (for Compact Linear Collider). This would involve using one beam to accelerate another one, and is both much more ambitious and much less well developed than the ILC technology. CERN hopes to have done enough research and development on

CLIC by 2010 to be able to decide whether to go ahead with a full design. Very optimistically, such a design effort would take five years or so, construction at least another five, so CLIC could not be completed until sometime after 2020. The ILC design could be constructed five years earlier, allowing its period of operation to overlap that of the LHC. Physicists would like to see such an overlap, since it is conceivable that something discovered at a linear collider could also be studied at the LHC, but might require a change in the experimental setup there. It now appears that the most likely course of events is that a decision to go ahead with a linear collider will be put off until at least 2009–2010, at which point some results from the LHC may be available. If the LHC finds Higgs or supersymmetric particles of low enough mass (these are postulated but never yet observed particles, whose significance will be explained in a later chapter) to be studied by the lower-energy ILC design, perhaps some way will be found to fund it. If no such particles are found, that would argue for waiting for the more ambitious, higher-energy CLIC design to be ready.

One other, more exotic, idea for a new collider has been under study in recent years. This would be a muon–antimuon collider at energies up to 4 TeV. Muons are much heavier particles than electrons, so there would be no synchrotron radiation problem, but they are unstable and decay to electrons and neutrinos in 2.2 microseconds. This lifetime is long enough for them to be stored for about five thousand revolutions in a 6-km storage ring. The main technical problem would be finding a way to achieve a useful luminosity in such a collider, but there is one other fundamental difficulty. As the muons decayed, an intense beam of neutrinos would be produced, one with enough energy and intensity to produce a radiation hazard. Neutrinos interact so weakly that most of them will pass all the way through the planet unaffected, so there is no way to shield against them. If such an accelerator were constructed 1km underground at Fermilab, neutrinos would emerge from the ground quite a distance away due to the curvature of the earth, producing a measurable radiation problem somewhere in the Chicago area.

Because of this and other problems, the muon storage ring designs are considered unlikely to lead to a viable collider anytime soon, but instead would make excellent neutrino factories. A neutrino factory would be an accelerator specifically designed to produce a controllable intense neutrino beam for the purpose of studying the properties of neutrinos. There are already several experiments in place at various locations around the world designed to study neutrinos, including the MiniBoone and NUMI/MINOS experiments previously mentioned. Typically, such experiments involve detectors located deep underground and look for collisions caused by neutrinos coming from the sun, from nuclear reactors, or from an accelerator (placed nearby or far away). This is one area of high-energy physics that does not necessarily require extremely expensive accelerators and detectors, and as such, its future is more assured than that of other kinds of experiments that physicists would like to be able to perform.

During much of the twentieth century, high-energy physics benefited from continual improvements in technology, which led to a long-lasting exponential rise in available energies. This phenomenon was somewhat like that of Moore's law in microelectronics, which says that the density of integrated circuits will double every 18 months, something that has had and will continue to have revolutionary implications for human society. Unfortunately for particle physicists, this period has come to an end. Without some unexpected revolutionary new technology, there is no chance that twenty-first-century accelerators will be able to continue the exponential rise that characterized the twentieth century. Particle theorists and experimentalists have entered a new period during which they must learn to deal with this difficult new reality.

Further reading

There are several excellent popular histories of particle physics that cover in detail the experimental side of the subject. Two of these are *The Particle Explosion*,[4] by Close, Sutton, and Marten, and *Discovery of Subatomic Particles*,[5] by Steven Weinberg.

Some other books and sources of information on the topics of this chapter are:

An influential recent book by the historian of science Peter Galison on the "material culture" of particle physics is *Image & Logic*.[6]

An anthropologist's take on the scientific community at SLAC can be found in Sharon Traweek's *Beamtimes and Lifetimes: The World of High Energy Physicists.*[7]

Up-to-date information on the performance of the Tevatron at Fermilab is available online at http://www.fnal.gov/pub/now.

Construction progress of the LHC can be followed at http://lhc.web.cern.ch/lhc.

For the latest news about the International Linear Collider project, see their website: http://www.linearcollider.org.

CERN Courier is a monthly magazine covering the latest news in particle physics. It is available online at http://www.cerncourier.com.

A website devoted to recent news and resources about particle physics is a http://www.interactions.org.

Symmetry is a monthly joint publication of SLAC and Fermilab, and is available at http://www.symmetrymag.org.

The High Energy Physics Advisory Panel provides advice to the Department of Energy and the National Science Foundation about the United States particle physics program. It meets about three times a year and many of the presentations to it are available online at http://www.science.doe.gov/hep/agenda.shtm.

Quantum Theory

Experimental data derived using the experimental techniques described in the last chapter have made possible the discovery of an extremely successful theoretical model for the physics of elementary particles now known as the "standard model." This chapter and the three following it will explain some of the concepts that go into the standard model, together with some of the history of their discovery.

In order to give a version of this story of reasonable length, several gross simplifications have been made. In particular, often the name of only one person is mentioned in connection with a discovery, although typically several people did work of a similar kind around the same time. This is also a Whiggish history: everything that didn't end up part of the current picture of things has been ruthlessly suppressed. Finally, while the last chapter began by emphasizing the role of experimentalists and their machines, now it's the theorist's turn, and the role of experiment will be minimized or ignored. The complex interaction of theory and experiment is just too intricate a tale to address in a few pages.

The standard model is a type of quantum field theory, of a kind known variously as a "Yang–Mills" or "nonabelian gauge" theory. The next few chapters will be concerned with quantum theory in general, quantum field theory, gauge theories, and finally the standard model itself.

QUANTUM THEORY AND ITS HISTORY

Up until the end of the nineteenth century, theoretical physics was about what we now usually refer to as classical physics. One part of this goes back to the seventeenth century and Isaac Newton's laws for the motion of particles experiencing a force. This subject is now known as Newtonian or classical mechanics. In classical mechanics, particles are characterized by a mass and two vectors that evolve in time: one vector gives the position of the particle, the other its momentum. Given a force acting on the particle, the evolution in time of the position and momentum vectors is determined by solving a differential equation given by Newton's second law. Newton developed both classical mechanics and the differential and integral calculus at the same time. Calculus is the language that allows the precise expression of the ideas of classical mechanics; without it, Newton's ideas about physics could not be usefully formulated.

By the latter part of the nineteenth century, the electromagnetic force that acts between charged particles had begun to be well understood. A new language for this was required, and the subject is now known as classical electrodynamics. Classical electrodynamics adds a new element, the electromagnetic field, to the particles of classical mechanics. In general, a field is something that exists throughout an extended area of space, not just at a point, like a particle. The electromagnetic field at a given time is described by two vectors for every point in space. One vector describes the electric field, the other the magnetic field. In 1864, James Clerk Maxwell formulated the Maxwell equations, a system of differential equations whose solution determines how the electric and magnetic field vectors interact with charged particles and evolve in time. The solutions to Maxwell's equations can be quite complicated in the presence of charged matter, but in a vacuum the solutions are quite simple and correspond to wavelike variations in the electromagnetic field. These are light waves: they can have arbitrary frequency, and they travel at the speed of light. Remarkably, Maxwell's equations manage to describe simultaneously two seemingly different sorts of things: the interactions of charged particles and the phenomenon of light.

Around the beginning of the twentieth century, it became clear to many physicists that the explanation of certain sorts of physical phenomena required ideas that went beyond those of classical mechanics and electrodynamics. In the cases of blackbody radiation for Max Planck (1900) and the photoelectric effect for Albert Einstein (1905), they found it necessary to invoke the idea that electromagnetic waves are "quantized." This hypothetical quantization postulated that for an electromagnetic wave of a given frequency, the energy carried by the wave could not be arbitrarily small, but had to be an integer multiple of a fixed energy: the "quantum." In 1913, the Danish physicist Niels Bohr discovered that he could predict which energies of light would be absorbed and emitted by hydrogen atoms by invoking a quantization principle for the orbits of electrons about the nucleus of the atom. Bohr's calculations were rather ad hoc, and did not constitute a consistent theory, but their close agreement with experiment made it clear that he was on the right track.

The basic ideas for a consistent theory of the quantum were worked out over a remarkably short period of time after initial breakthroughs by Werner Heisenberg and Erwin Schrödinger in 1925 and 1926. This set of ideas came to be known as quantum mechanics and revolutionized scientists' picture of the physical world. These ideas remain to this day at the very core of theoretical physics. Quantum mechanics subsumes classical mechanics, which now appears merely as a limiting case approximately valid under certain conditions. The picture of the world provided by quantum mechanics is completely alien to that of our everyday experience, which is in a realm of distance and energy scales where classical mechanics is highly accurate. In quantum mechanics, the state of the world at a given time is given not by a bunch of particles, each with a well-defined position and momentum vector, but by a mathematical abstraction, a vector specified not by three spatial coordinates, but by an infinite number of coordinates. To make things even harder to visualize, these coordinates are complex numbers, not real numbers. This infinite-dimensional "state space" of vectors is called Hilbert space, after the German mathematician David Hilbert.

While states in quantum mechanics can abstractly be thought of as vectors in Hilbert space, a somewhat more concrete alternative is to think of them as "wave functions": complex-valued functions of the space and time coordinates. Such functions behave like vectors in that one can add them together or scale them by a constant. Depending on context, it can be useful to think of quantum-mechanical states either as abstract vectors in Hilbert space or as wave functions, but these two points of view are completely equivalent.

The fundamental conceptual and mathematical structure of quantum mechanics is extremely simple and has two components:

1. At every moment in time, the state of the world is completely described by a vector in Hilbert space.
2. Observable quantities correspond to operators on Hilbert space. The most important of these is called the Hamiltonian operator, and it tells one how the state vector changes with time.

Note that this fundamental structure is not probabilistic, but just as deterministic as classical mechanics. If one knows precisely the state vector at a given time, the Hamiltonian operator tells one precisely what it will be at all later times (the explicit equation one solves to do this is Schrödinger's equation). Probability comes in because one doesn't directly measure the state vector, but instead measures the values of certain classical observables (things such as position, momentum, and energy). But most states don't correspond to well-defined values of these observables. The only states with well-defined values of the classical observables are the special ones that are called "eigenstates" of the corresponding operator. This means that the operator acts on the state in a very simple way, just by multiplying it by some real number, and this real number is then the value of the classical observable. When one does a measurement, one does not directly measure the state of a system, but interacts with it in some way that puts it into one of those special states that does have well-defined values for the particular classical observables that one is measuring. It is at this point that probability comes in, and one can predict only what the probabilities will be for finding various values for the different possible classical observables.

This disjunction between the classical observables that one measures and the underlying conceptual structure of the theory is what leads to all sorts of results that violate our intuitions based on classical physics. One famous example is Heisenberg's uncertainty principle, which originates in the fact that there is no vector in Hilbert space that describes a particle with a well-defined position and momentum. The position and momentum operators do not commute (meaning that applying them in different orders gives different results), so state vectors that are "eigenstates" for both position and momentum do not exist. If one considers the state vector corresponding to a particle at a fixed position, it contains no information at all about the momentum of the particle. Similarly, the state vector for a particle with a precisely known momentum contains no information about where the particle actually is.

The new theory required some getting used to, and one could not appeal to physical intuition to justify it. This lack of *Anschaulichkeit* (visualizability) was of great concern to the founders of the theory. On the other hand, the theory had tremendous explanatory power, giving detailed and precise predictions for many aspects of atomic spectra and a huge array of other physical phenomena that had resisted any explanation in the language of classical mechanics and electrodynamics. The mathematics involved was new to physicists, but well known to Hilbert and many others of Heisenberg's mathematical colleagues at Göttingen. In the fall of 1925, a few months after his initial breakthrough, Heisenberg was in Göttingen, working out, with his physicist colleagues Max Born and Pascual Jordan, an initial version of quantum mechanics that is often called matrix mechanics. Born was a Göttingen mathematics PhD who had changed fields to theoretical physics, and Jordan was a student of his who had toyed with becoming a mathematician. At this time, Heisenberg wrote to Wolfgang Pauli,

> I've taken a lot of trouble to make the work physical, and I'm relatively content with it. But I'm still pretty unhappy with the theory as a whole and I was delighted that you were completely on my side about mathematics and physics. Here I'm in an environment that thinks and feels exactly the opposite way and I don't know whether I'm just too stupid to understand the mathematics. Göttingen is divided into two camps: one, which speaks, like Hilbert (and Weyl, in a letter to Jordan), of the great success that will follow the development of matrix calculations in physics; the other, which like [physicist James] Franck, maintains that the matrices will never be understood.[1]

One of the intellectual leaders of the new quantum mechanics was the brilliant young English physicist Paul Adrien Maurice Dirac. In September 1925, aged 23 and while still a PhD student at Cambridge, he began studying Heisenberg's first paper on his matrix mechanics version of quantum mechanics, which had just appeared. He soon found a beautiful relationship between the mathematics of Heisenberg's matrices and the mathematics of classical mechanics. This had far-reaching implications, and was to become one of the cornerstones of the new quantum mechanics.

Dirac was known both for the mathematical sophistication of his ideas and for being a man of few words. On a visit to the University of Wisconsin in 1929, a local journalist came to interview him, leading to the following exchange:

> And now I want to ask you something more: They tell me that you and Einstein are the only two real sure-enough high-brows and the only ones who can really understand each other. I won't ask you if this is straight stuff for I know that you are too modest to admit it. But I want to know this—Do you ever run across a fellow that even you can't understand?
>
> Yes.
>
> This will make a great reading for the boys down at the office. Do you mind releasing to me who he is?
>
> Weyl.[2]

The "Weyl" that Dirac and Heisenberg were having so much trouble understanding was Hermann Weyl, one of the leading mathematicians of the day. He had begun his career at Göttingen, where he was a student of Hilbert and then held his first teaching position. In 1925, he was professor of mathematics in Zurich, the same city where the physicist Schrödinger was developing the version of quantum mechanics known as wave mechanics. Weyl was Schrödinger's closest friend when he came to Zurich, and the two shared a common interest in the night life of the city.[3]

Schrödinger's revolutionary work on the wave mechanics version of quantum mechanics came together in the last weeks of December 1925. He was staying in an inn up in the mountains at Arosa with a girlfriend from Vienna, one whose identity remains a mystery to this day. He was working on what was to become known as the "Schrödinger equation," and trying to understand the mathematics of the solutions of this equation for the crucial case of the hydrogen atom. When he got back to Zurich, he consulted Weyl, who was an expert on this kind of equation, and explained to him what the general properties of its solutions were. In his first paper on quantum mechanics,[4] Schrödinger explicitly thanks Weyl for his help.

Weyl later commented on this period by remarking that Schrödinger "did his great work during a late erotic outburst in his life."[5] Schrödinger was married, but was "convinced that Bourgeois marriage, while essential for a comfortable life, is incompatible with romantic love."[6] His wife, Annemarie, presumably was not too concerned about his spending time in the mountains with his girlfriend, since she was Weyl's lover at the time. The relationship between mathematics and physics at the birth of quantum mechanics was an intimate one indeed, with Schrödinger and Weyl sharing much more than purely intellectual interests.

Most histories of quantum mechanics are written by physicists and pay little if any attention to the mathematician Weyl's role in this story. Besides his close connection to Schrödinger and the Schrödinger equation, he was in communication with many of the other main figures in the story and well aware of what they were discovering. What is not clear is whether they understood much at all of what he was telling them about a mathematician's point of view on the new theory they were investigating. A significant digression into mathematics is required in order to give some idea of what Weyl's perspective was.

MATHEMATICAL DIGRESSION: SYMMETRY GROUPS AND REPRESENTATIONS

What was it that Weyl was telling them that Dirac and Heisenberg found interesting and yet very difficult to understand? This chapter will attempt to give a general idea of the mathematics that is at issue, but keep in mind that a real understanding of this subject cannot possibly be had from reading these few pages. Just as quantum mechanics is a difficult and counterintuitive subject to understand and yet is at the center of modern physics, this sort of mathematics takes a lot of effort to develop intuitions about, and is one of the central subjects of modern mathematics. Like quantum mechanics, it is very much a product of the twentieth century, and the history of the two subjects is significantly intertwined.

Many readers may find the material in this section hard to follow, so here is a short summary. When there are transformations of a physical system that do not change the laws of physics governing the system, these transformations are said to be "symmetries" of the physical system. An important example is translation in space or time: for most experiments, if the experimenter picks up the experimental apparatus and moves it a bit in any direction, or waits a little while to do the experiment, this kind of transformation of the system will not change the result of the experiment. A set of symmetry transformations is an example of an abstract structure that mathematicians call a group. The groups that turn out to be important in physics include not only the translations, but also rotations and other groups defined using complex numbers with names like U(1), SU(2), and SU(3).

Given a physical system described using quantum mechanics, if it has a group of symmetry transformations, the quantum-mechanical Hilbert space of states is an example of a mathematical structure called a representation of the group. Very roughly, the group is the set of symmetries, and the representation is the thing that experiences symmetry transformations. Some of the most basic aspects of physics follow from looking at symmetries. The symmetry under translations in space implies the conservation of momentum, while symmetry under translation in time implies the conservation of energy. The relationship between these conservation laws and the symmetry transformations is much more direct in quantum theory than it is in classical mechanics.

Groups, Representations, and Quantum Mechanics

During 1925 and 1926, Weyl was doing what was probably the most important work of his career, working out a significant part of what is now known as the representation theory of Lie groups, a set of terms that this chapter will try to explain. To a mathematician, a group is just a set of abstract elements that one can manipulate by multiplying them together, with the crucial feature that there is one distinguished element called the identity element, and each element is paired with an inverse element. Multiplying an element by its inverse gives the identity. While one can consider these abstract groups in and of themselves, the more interesting part of the subject is representation theory, where one represents each abstract element in the group as a transformation of something else. These transformations are meant to be symmetry transformations; the transformation does not completely change things, but it leaves invariant some structure or other that one is interested in. Multiplying together two group elements just corresponds to composing the two corresponding symmetry transformations, performing one after the other. Taking the inverse of a group element corresponds to undoing the symmetry transformation.

When one has a representation of a group, one can think of the group as a symmetry group: each of its elements is a symmetry transformation of something. A simple but nontrivial example of this is to consider a group of two elements, represented as two transformations of three-dimensional space. The first transformation is the trivial one: just don't change anything. This is called the identity transformation. The second transformation is the interesting and nontrivial one. It is given by reflecting everything in a mirror. Doing two mirror reflections in a row gives back exactly the original situation. The mathematics of the group involved here is very simple: there are just two elements. One of them is the "identity" (multiplication by it doesn't do anything); the other is such that multiplying it by itself gives the identity element. The representation is a lot more interesting than the group itself. It brings in the concept of three-dimensional space and an interesting transformation (mirror reflection) that can be performed on it.

The mathematics of groups consisting of a finite number of elements goes back to the French mathematician Evariste Galois, who, in the years before his early death in a duel at the age of 21 in 1832, used them to show the impossibility of finding a formula for the solutions of polynomial equations of degree five or higher. These finite groups and their representations were studied during the late nineteenth century, a period that also saw the formulation in 1873 by Sophus Lie of the definition of a Lie group. A Lie group is also called a continuous group, since it consists of an infinite number of elements continuously connected together. It was the representation theory of these groups that Weyl was studying.

A simple example of a Lie group together with a representation is that of the group of rotations of the two-dimensional plane. Given a two-dimensional plane with chosen central point, one can imagine rotating the plane by a given angle about the central point. This is a symmetry transformation of the plane. The thing that is invariant is the distance between a point on the plane and the central point. This is the same before and after the rotation. One can actually define rotations of the plane as precisely those transformations that leave invariant the distance to the central point. There is an infinity of these transformations, but they can all be parameterized by a single number, the angle of rotation.

If one thinks of the plane as the complex plane (the plane whose two coordinates label the real and imaginary parts of a complex number), then the rotations can be thought of as corresponding not just to angles, but to a complex number of length one. If one multiplies all points in the complex plane by a given complex number of unit length, one gets the corresponding rotation (this is a simple exercise in manipulating complex numbers). As a result, the group of rotations in the complex plane is often called the "unitary group of transformations of one complex variable," and written U(1).

This is a very specific representation of the group U(1), the representation as transformations of the complex plane, but one can also ask what other representations this group might have. It turns out that this question leads to the subject of Fourier analysis, which began with the work of the French mathematician Joseph Fourier in the very early nineteenth century. Fourier analysis is the study of how to decompose periodic, wavelike phenomena into a fundamental frequency and its higher harmonics. It is also sometimes referred to as harmonic analysis. To go into the details of the relationship between harmonic analysis and representation theory would require a longer discussion, but one thing to note is that the transformation of rotation by an angle is formally similar to the transformation of a wave by changing its phase. Given an initial wave, if one imagines copying it and then making the copy more and more out of phase

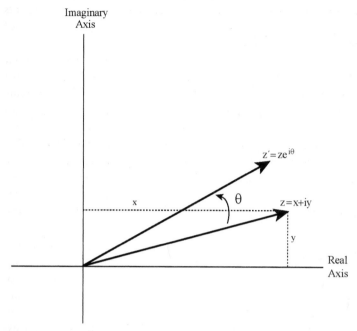

Rotation by an angle θ in a complex plane.

with the initial wave, sooner or later one will get back to where one started, in phase with the initial wave. This sequence of transformations of the phase of a wave is much like the sequence of rotations of a plane as one increases the angle of rotation from 0 to 360 degrees. Because of this analogy, U(1) symmetry transformations are often called phase transformations.

The example of the Lie group U(1) is an extremely simple one, and an important reason for its simplicity is that the rotations of a plane are what a mathematician calls commutative transformations: when two of them are performed sequentially, it doesn't matter which is done first and which is done second. When one goes to higher dimensions, things immediately get much more complicated and much more interesting. In three dimensions, one can again think about the group of rotations, represented as rotations of three-dimensional space. These rotations are the transformations of

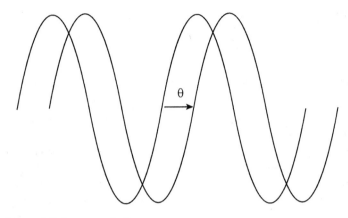

Phase shift by an angle θ.

three-dimensional space that leave the distance to a central point unchanged, the same definition as in two dimensions. If one performs two rotations about two different axes, the overall rotation one gets depends on the order of the two rotations. Thus the group of rotations in three dimensions is what is called a noncommutative, or nonabelian, group. It was the theory of representations of this kind of group that Weyl was working on in 1925–1926.

Weyl's theory applies to many different kinds of groups of transformations in higher dimensions, but the simplest cases are those in which one takes the coordinates of these higher dimensions to be complex numbers. The case of the plane is the case of one complex number; the next case involves two complex numbers. Now one is in a situation where it is no longer possible to visualize what is going on. Two complex numbers correspond to four real numbers (two real parts and two imaginary parts), so these transformations are happening in four (real) dimensions. To make visualization even trickier, there is an additional subtle piece of geometrical structure that does not exist when one is dealing just with real numbers. In the complex plane, multiplication by the imaginary unit (square root of minus one) corresponds geometrically to a 90-degree counterclockwise rotation. In four dimensions, there is not just one possible axis of rotation as in two dimensions, but an infinity of axes about which one could imagine rotating counterclockwise by 90 degrees. The identification of the four dimensions with two complex numbers picks out one of these axes: it is the axis about which one rotates when one multiplies the two complex numbers by the imaginary unit. So, two complex dimensions have both one more real dimension than one can visualize, and in addition have an unvisualizable extra structure.

Despite this lack of visualizability, symmetry transformations of these higher complex dimensions can be easily written down algebraically. The formalism for doing this uses matrices of complex numbers; this is the formalism that Heisenberg was not so happy about. In general, if one has an arbitrary number N of complex numbers, one can define the group of unitary transformations of N (complex) variables and denote it by U(N). It turns out that it is a good idea to break these transformations into two parts: the part that just multiplies all of the N complex numbers by the same unit-length complex number (this part is a U(1) like before), and the rest. The second part is where all the complexity is, and it is given the name of special unitary transformations of N (complex) variables and denoted by SU(N). Part of Weyl's achievement consisted in a complete understanding of the representations of SU(N), for any N, no matter how large.

In the case $N = 1$, SU(1) is just the trivial group with one element. The first nontrivial case is that of SU(2). The symmetry group SU(2) has a very special role that brought it into play from the earliest days of quantum mechanics. It turns out, for not at all obvious reasons (but then, not much about the geometry of pairs of complex numbers is obvious), that the group SU(2) is very closely related to the group of rotations in three real dimensions. Rotations of three dimensions are extremely important, since they correspond to symmetry transformations of the three space dimensions of our real world, so physical systems can provide representations of this group. Mathematicians call the group of rotations of three-dimensional space the group of special orthogonal transformations of three (real) variables and denote this group by SO(3). The precise relation between SO(3) and SU(2) is that each rotation in three dimensions corresponds to two distinct elements of SU(2), so SU(2) is in some sense a doubled version of SO(3).

With any physical system one can associate a symmetry group, the group of symmetry transformations that one can perform on the system without changing any of the physics. For instance, in the case of atomic spectra, the physics does not change as one performs rotations about the center of the atom defined by the nucleus. In the laws of atomic physics there is no distinguished direction in space. All directions are equivalent, and any two different directions are related by a rotation transformation that takes one into the other. One says that the laws governing an atom have "rotational symmetry," and the group SO(3) of three-dimensional rotations is a symmetry group of the situation.

Long before quantum mechanics, it had been known that something very important happens in classical mechanics whenever one has a symmetry group. What happens in a classical mechanical system with symmetry is that one can define conserved quantities. These are combinations of the momentum and position that stay the same as the system evolves in time. Because of the nature of three-dimensional space, physical systems have two standard sorts of symmetries. The first is symmetry under translations: one can move a physical system around in any of the three possible directions without changing the laws of physics. The three conserved quantities in this case are the three components of the momentum vector. The fact that the components of the total momentum of a physical system do not change with time is one of the fundamental principles of Newtonian mechanics. The second standard symmetry is symmetry under the SO(3) group of rotations. This group is three-dimensional (there are three axes about which you can rotate), so there are three more conserved quantities. These are the three components of a vector called the angular momentum vector, and again it is a fundamental principle of mechanics that the components of the total angular momentum of an isolated system experiencing rotationally symmetric forces do not change.

If one thinks in terms of Einstein's special relativity, where the three dimensions of space and the one dimension of time are supposed to be considered together, there is one more kind of translation, translation in the time direction. The corresponding conserved quantity is the energy. One can also try to think about rotations in four dimensions, but this is a rather subtle business given that one of the directions, time, is different from the others.

The existence of these conserved quantities, together with the associated conservation laws that say that they don't change, is one of the most fundamental facts about any physical system. In studying virtually any sort of physics, identifying the energy, momentum, and angular momentum and using their associated conservation laws are often starting points of any analysis of what is going on. The fact that these conservation laws come about in classical mechanics because of the symmetries of the situation is known to most physicists as Noether's theorem, after the German mathematician Emmy Noether.

As a student, Noether spent one semester in Göttingen, where she attended Hilbert's lectures, but did most of her studies and completed her dissertation at Erlangen, since it was one place that would accept female students. She returned to Göttingen in 1915, but could not be given a formal position as an instructor until after the war, when political changes improved the legal status of women. In his memorial address after her death,[7] Weyl recalls how Hilbert argued her case at a faculty meeting against opposition from philologists and historians: "I do not see that the sex of the candidate is an argument against her admission as a Privatdozent. After all, we are a university, not a bathing establishment." Finally, Hilbert dealt with the problem by announcing a lecture course in his own name, and then having Noether give the lectures. Weyl was a visitor to Göttingen and lectured on representation theory during the winter semester of 1926–1927. Noether's interests were more algebraic, and she never directly worked on quantum mechanics, but she attended Weyl's lectures and would often discuss mathematics with him walking home afterward. When Weyl came to Göttingen in 1930, he noted that she was the major center of mathematical activity there. She finally left in 1933 to escape the Nazis, ending up teaching in the United States at Bryn Mawr until her death in 1935.

After finishing his work on the representation theory of Lie groups in 1926, Weyl turned his attention to the question of how to use representation theory as a tool to work out the implications in quantum mechanics of various symmetry groups. His book on the subject, *Gruppentheorie und Quantenmechanik,*[8] was published in 1928 and had a great influence on both mathematicians and physicists. It turns out that while the formalism of quantum mechanics violates our everyday intuitions, it fits very closely with group representation theory. The analysis of the implications of the existence of a symmetry group for a physical system is actually quite a bit simpler in quantum mechanics than in classical mechanics.

The languages of physics and mathematics had already grown apart by the 1920s, and many physicists found Weyl's writing style hard to follow. The Hungarian physicist Eugene Wigner, who was in Göttingen during 1927 and 1928, described the situation as follows:

> So Hermann Weyl thought very clearly, and his textbook *Group The-ory and Quantum Mechanics,* first published in 1928, had become the standard text in that field. Those who understood it saw in it a rigorous beauty. But Weyl did not write clearly, and so most physicists did not understand his book. Young students especially found the book awfully dense. For all his brilliance and good intentions, Hermann Weyl had discouraged a fair number of physicists from studying group theory.[9]

Wigner also describes Pauli as giving the label *die Gruppenpest* (the plague of group theory) to the increasing use of group theory in quantum mechanics, and this term was popular among physicists for many years. Wigner himself did a lot to improve the situation, writing a book entitled *Group Theory and Its Application to the Quantum Mechanics of Atomic Spectra* in 1931. It explained groups and representations in a language closer to that with which physicists were familiar.

The response of the postwar generation of physicists to Weyl's book was similar. The physicist C. N. Yang, whose work played an important role in the standard model, remembers:

> His [Weyl's] book was very famous, and was recognized as profound. Almost every theoretical physicist born before 1935 has a copy of it on his bookshelves. But very few read it: Most are not accustomed to Weyl's concentration on the structural aspects of physics and feel un-

comfortable with his emphasis on concepts. The book was just too abstract for most physicists.[10]

Given any physical system described by quantum mechanics, if the system has a symmetry group, then the Hilbert space of its states is precisely the sort of representation of the symmetry group that Weyl had been studying. So the Hilbert space may be something that seems not at all natural to our classical physical intuition, but it is exactly what a mathematician studying group representations expects to find. Recall that the founders of quantum mechanics had discovered that they had to stop thinking in terms of the momentum vector and the position vector of classical mechanics, but instead had to start thinking in terms of operators on the Hilbert space. The operator on the Hilbert space corresponding to a momentum vector is precisely the operator on the Hilbert space that implements infinitesimal translations in the direction of the momentum vector. Similarly, the operator corresponding to an angular momentum vector is precisely the one that implements an infinitesimal rotation about the axis given by that vector. In quantum mechanics, the way in which conserved quantities come from symmetries is much more direct than in classical mechanics. Instead of a conserved quantity corresponding to a symmetry transformation, one has simply the operator that implements an infinitesimal symmetry transformation on the Hilbert space of state vectors.

There are some new symmetries in quantum mechanics that do not exist in classical mechanics. If one multiplies all state vectors in Hilbert space by the same unit-length complex number, or in other words, just changes the overall phase of the wave function, the observed physics is not supposed to change. So a quantum-mechanical system naturally has a U(1) symmetry. To this symmetry there is a corresponding conserved quantity, the electric charge. The conservation law in this case is just the well-known principle of conservation of charge. Thus, another one of the most fundamental aspects of physics, charge conservation, now appears as a result of a simple symmetry principle.

Of the standard symmetries of physical systems, the one that requires the most nontrivial mathematics to analyze is rotational symmetry. The representations of the rotational symmetry group SO(3) can all in some sense be built out of one so-called fundamental representation: the one that comes from just rotating three-dimensional vectors in three-dimensional space. The fundamental representation of SO(3) is this representation on three-dimensional vectors, with these vectors moving under rotations in the obvious way that an arrow would rotate about the origin. In 1924, just before quantum mechanics burst on the scene, Pauli had suggested that many aspects of atomic spectra could be understood if the electron had a peculiar double-valued nature. Thinking in terms of quantized states, the electron seemed to have twice as many states as one would naively expect. Once the relation between quantum mechanics and representation theory was understood, it became clear that this happens because the Hilbert space for an electron is not a representation of the SO(3) rotational symmetry, but is a representation of the related group SU(2). Recall that SU(2) is a sort of doubled version of SO(3), and it is the group of special unitary transformations on two complex variables. The representation of SU(2) that defines the group is this representation on sets of two complex numbers. This is exactly what is needed to describe an electron, together with its properties as a representation of the symmetry group of rotations. The two complex numbers instead of one account for Pauli's doubling of the number of states.

The SU(2) transformation properties of a particle have become known as the particle's spin. This term comes from the idea that one could think of the particle as a spinning particle, spinning on some axis and thus carrying some angular momentum. This idea is inherently inconsistent for a number of reasons. While the spin is a quantized version of the angular momentum, there is no well-defined axis of rotation or speed of rotation. Spin is an inherently quantum-mechanical notion, one that fits in precisely with the representation theory of the symmetry group SU(2), but has no consistent interpretation in terms of classical physics.

The classification of the different kinds of representations of SU(2) is very simple and goes as follows:

- There is a trivial representation, which is said to have spin 0. In this representation, no matter what rotation one performs, there is no effect on the representation.

- There is the fundamental representation of SU(2) on pairs of complex numbers. This is said to have spin one-half and is what is used to describe the state of an electron.
- There is the representation of SU(2) by the usual three-dimensional rotations acting on three-dimensional vectors. This is called the spin-one representation and is used to describe photons.

There are higher-dimensional representations, one for each half-integer. The only other one known to have possible significance for elementary particles is the spin-two representation, which may be needed to describe gravitons, the hypothetical quanta of the gravitational field.

This has been a significantly oversimplified description of representation theory and its connection to quantum mechanics. One important complication is that the Hilbert spaces that appear in quantum mechanics are actually infinite-dimensional. Weyl's work on representation theory gave complete answers only for finite-dimensional representations. Information about finite-dimensional representations was a very important part of what was needed to understand the case of infinite-dimensional Hilbert space, but it wasn't everything. Quantum mechanics had raised many basic questions about representations that mathematicians could not answer. These questions were an important stimulus for new mathematical work from the 1930s on. While mathematicians went to work studying some of these issues, physicists mostly had turned their attention in another direction, one that will be the subject of the next chapter.

FURTHER READING

There are many textbooks on quantum mechanics available at a wide range of levels. At least for concision, it is hard to surpass Dirac's classic *The Principles of Quantum Mechanics*.[11] A good recent nontechnical introduction to the subject is *The New Quantum Universe*,[12] by Hey and Walters. An older one is *The Story of Quantum Mechanics*,[13] by Guillemin. An encyclopedic reference for the history of quantum theory is the six-volume set *The Historical Development of Quantum Theory*,[14] by Mehra and Rechenberg.

There are quite a few popular books by physicists explaining the importance of the notion of symmetry in physics, although some of them avoid discussing the concept of a group representation. Some good examples are *Fearful Symmetry*,[15] by Zee, *Longing for the Harmonies*,[16] by Wilczek and Devine, and *The Equation that Couldn't Be Solved*,[17] by Livio. Wigner's collection of essays *Symmetries and Reflections*[18] treats many topics related to quantum mechanics and mathematics. See also Weyl's *Symmetry*,[19] a nontechnical book written late in his life.

Two good recent mathematics textbooks on the representation theory of Lie groups are *Lie Groups, Lie Algebras and Representations*,[20] by Hall, *Representations of Finite and Compact Groups*,[21] by Simon, and *Lie Groups*,[22] by Rossman. A detailed history of the subject up to Weyl's work is *Emergence of the Theory of Lie Group*,[23] by Hawkins.

For an introduction to spinors and their geometry, an excellent reference is Penrose's recent *The Road to Reality*.[24]

4

Quantum Field Theory

In the years immediately after 1925, quantum mechanics was extended to explain the behavior of a wide variety of different physical systems. An early extension was to the case of the electromagnetic field. Here, the general principles of quantum mechanics could be used to justify the quantization of electromagnetic waves first discovered by Planck and Einstein. A classical field theory (such as electromagnetism) treated according to the principles of quantum mechanics became known as a quantum field theory. The quantized excitations of the electromagnetic field were called photons and had the confusing property of wave–particle duality. This term referred to the fact that in some situations, such as those in which classical physics was approximately valid, it was easiest to understand the behavior of the quantized electromagnetic field in terms of waves, but in others it was best to think in terms of particles, the photons.

When applied to particles such as electrons, quantum mechanics was limited to dealing with a fixed and finite number of particles. Physical systems with different numbers of electrons corresponded to different Hilbert spaces. While this was adequate for understanding the part of atomic physics that is concerned with a fixed number of electrons moving around an atomic nucleus, at high enough energy, situations occur in which the number of electrons will change. The well-known equation $E = mc^2$ of Einstein's special relativity gives the energy contained in a massive particle at rest. If at least twice that energy is available, say in a high-energy photon, the photon can transform into an electron–positron pair, where the positron is the antiparticle of the electron. This kind of change in the number of particles cannot be accommodated in quantum mechanics, but can be accounted for in a quantum field theory.

Even if one is dealing purely with situations in which the overall energy is too low to create particle–antiparticle pairs, combining the principles of special relativity and quantum mechanics indicates that there is an inherent inconsistency in trying to stick to a fixed number of particles. Recall that in quantum mechanics, Heisenberg's uncertainty principle says that as one tries to locate the position of a particle to greater and greater accuracy, its momentum becomes undetermined and could be measured as being larger and larger. At some point the unknown momentum can be so large that the corresponding kinetic energy is large enough to produce particle–antiparticle pairs. In other words, if one tries to confine a particle to a smaller and smaller box, at some point one not only doesn't know what its momentum is, one also doesn't know how many particles are in the box.

The quantum field theory of the electromagnetic field was a consistent quantum theory in which particles (the photons) could be created or annihilated. In order to be able to deal with the phenomenon of electrons being created or annihilated, it was natural to conjecture that not just photons, but all particles, electrons included, were the quantized excitations of some quantum field theory. There should be an "electron field" such that when one makes a quantum field theory out of it, one gets electrons instead of photons.

Central to Schrödinger's wave mechanics version of quantum mechanics was an equation for the time evolution of the wave function describing the state of the system. Unfortunately, Schrödinger's equation had a major problem that he had been aware of from the beginning. It was inconsistent with the principles of special relativity. Schrödinger had actually first written down an equation that was consistent with special relativity, but quickly abandoned it when he found that it gave results not in accordance with experiment. The equation that he published and to which his name was given agreed with special relativity only in the approximation that all particles involved were moving at much less than the speed of light. This approximation was good enough to deal with a large number of questions about atomic spectra, but was guaranteed to become problematic at energies at which particle–antiparticle pairs could be produced. Finally, another problem for the equation was that it had to be modified by adding special terms to account for Pauli's discovery that electrons carry the surprising property of spin.

Late in 1927, Dirac discovered a remarkable equation that is now known as the Dirac equation. Dirac's equation is analogous to Schrödinger's, but is consistent with special relativity, and it automatically explains the spin of the electron. While some of its solutions describe electrons, other solutions describe positrons, the antiparticle of the electron. These positrons were experimentally observed a few years later in 1932. The explanatory power of the Dirac equation is hard to overemphasize. In one fell swoop it solved problems about atomic spectra by properly explaining electron spin, reconciled quantum mechanics with special relativity, and predicted a new particle that was soon found. Surprisingly, a crucial part of the story was Dirac's rediscovery of an algebraic gadget called a Clifford algebra, named after the English mathematician William Clifford, who first wrote about them in 1879. When physicists talk about the importance of beauty and elegance in their theories, the Dirac equation is often what they have in mind. Its combination of great simplicity and surprising new ideas, together with its ability to both explain previously mysterious phenomena and predict new ones, makes it a paradigm for any mathematically inclined theorist.

Mathematicians were much slower to appreciate the Dirac equation, and it had little impact on mathematics at the time of its discovery. In contrast to its impact on physicists, the equation did not immediately answer any questions that mathematicians had been thinking about. This began to change in the early 1960s, when the British mathematician Michael Atiyah and his American colleague Isadore Singer rediscovered the equation for themselves in their work on what is called the Atiyah–Singer index theorem, one of the most important results in mathematics of the latter half of the twentieth century.

With the Dirac equation in place with its unambiguous prediction of the necessity of dealing with the production of electron–positron pairs, work on a quantum field theory of the electron field began in earnest. By the end of 1929, Jordan, Pauli, and Heisenberg had written down a full quantum field theory of electrons and the electromagnetic field. This theory was to become known as quantum electrodynamics and acquired the acronym QED. The year 1929 marked the end of the most spectacular four-year period in the history of physics. During this short time the subject had gone from some incoherent and ad hoc ideas about atomic physics to the formulation of a complete theory that survives more or less unchanged to this day, and is the foundation of the modern picture of the physical world.

From this time on, progress slowed dramatically for several different reasons. The rise of the Nazis turned many German mathematicians and physicists into refugees and destroyed the great German scientific communities such as the one at Göttingen. Weyl, Schrödinger, and Born left in 1933 for the United States or England. Heisenberg and Jordan demonstrated that scientific and moral intelligence are two very different things by collaborating with the Nazis. During the war years of 1939–1945, many physicists on both sides were involved in one way or another in the war effort. Heisenberg led the German effort to develop a nuclear weapon, luckily in the process showing that brilliant theoretical physicists are not necessarily competent to run experimental physics and engineering projects.

The age of the particle accelerator had begun in earnest with the first cyclotrons in 1931, and many theorists were kept busy trying to interpret the new results about the physics of the atomic nucleus being produced by these machines and by cosmic ray experiments. New particles were being discovered: the neutron in 1932, the muon in 1937, and the pion in 1947, adding baffling new mysteries to consider. The new quantum field theory QED seemed to have nothing useful to say about any of this. To some extent it could be extended to describe the new particles (for instance, the muon behaves just like a more massive electron), but many of these new particles seemed to have properties that could not be accounted for within QED.

Progress on QED had come to a halt for another fundamental reason purely internal to the theory. QED is a theory for which no exact solution is known, a situation that continues to this day. The main calculational method available is something called a perturbation expansion. The idea here is essentially the same as that of a power series expansion in calculus. There, if one doesn't know how to calculate the values of a function, one works with a sequence of polynomials that is supposed to be a better and better approximation to the function as one takes polynomials of higher and higher degree. Such a power series expansion starts with a zeroth-order term, which is just the value of the unknown function at some point where one does know how to calculate it. In the case of the perturbation expansion for QED, the zeroth-order term is the theory with the strength of the interaction between the electron and the electromagnetic field set to zero. In this approximation, QED is a theory of free particles and is exactly solvable. One has electrons, positrons, and photons, but they pass through each other with no effect. The next term in the perturbation expansion, the first-order term, is also exactly calculable, and in this approximation the theory looks a lot like the real world. Most interesting physical phenomena appear in this first-order approximation to the real QED, and the approximated theory agrees well with experimental results.

So far so good. Unfortunately, it was soon discovered that calculations of the second- and higher-order terms in the perturbation theory for QED appeared to make no sense. Each time one goes to the next-higher order in this kind of calculation, one picks up a factor known as the fine-structure constant and denoted by α, which has a value of roughly $\frac{1}{137}$. So one expects the second-order terms to be about 137 times smaller than the first-order terms. Instead, what happens is that the second-order terms appear to be infinite. Something has gone seriously wrong, for the second-order calculation, instead of giving a better approximation to the real world, gives something that makes no sense.

This situation was profoundly discouraging, and led to the widespread belief that there was something fundamentally wrong with QED as a theory. Many theorists gave up on QED and went to work on other problems. A few continued to try to make sense of the perturbation expansion, and during the 1930s it became clear that the source of the problem was something called "renormalization." While in the first-order calculation the parameters in the theory that set the charge and mass of an electron correspond exactly to what one observes, in the higher-order calculation these parameters have a nontrivial relation to the observed charge and mass. The conjectured solution to the problem of the infinities was that one needed to set up the calculation so that all results were expressed purely in terms of the observed charge and mass.

This was where things stood at the beginning of the war, and very little progress was made during the war years. After the war, a new generation of theoretical physicists began attacking the renormalization problem. The first success came in 1947, when Hans Bethe calculated a finite value for higher-order corrections to certain atomic energy levels. The first-order calculation of these levels predicts that two of them will be exactly the same, but earlier in that year, the experimentalist Willis Lamb had first measured a difference in the two energies, a difference that came to be known as the Lamb shift. Bethe's calculation of the difference of energy levels did not fully take into account the effects of special relativity, but it agreed fairly well with Lamb's experimental value. By 1949, Julian Schwinger, Richard Feynman, and Freeman Dyson in the United States and, completely independently, Sin-Itiro Tomonaga in Japan had been able to push through calculations in a properly renormalized perturbation expansion of QED and show that in principle these calculations could be done to arbitrarily high order.

Feynman came up with a graphical representation of the terms in the perturbation expansion, and his diagrammatic method allowed calculations to be performed relatively easily. To this day, all students of quantum field theory begin to study the subject by learning how to calculate these Feynman diagrams. Much lore and varied terminology has grown up around them. In the early 1960s, Harvard physicist Sidney Coleman coined the term "tadpole diagrams" to refer to Feynman diagrams that contain a line segment ending in a circle. When the editors of *Physical Review* objected to this terminology, he agreed to change it, but suggested "lollipop" or "sperm diagram" as an alternative terminology. They settled on "tadpole." Another kind of Feynman diagram is called a penguin diagram, something that confused me when I first learned about them, since they do not look much like penguins. The story behind this seems to be that particle theorist John Ellis and experimentalist Melissa Franklin were playing darts one evening in 1977 at CERN, and a bet was made that would require Ellis to insert the word "penguin" in his next research paper if he lost. He did lose but was having a lot of trouble figuring out how he would fulfill his obligation. Finally, "the answer came to him when one evening, leaving CERN, he dropped by to visit some friends where he smoked an illegal substance." While working on his paper later that night, "in a moment of revelation he saw that the diagrams looked like penguins."[1]

The measurement of the Lamb shift was made possible by technological innovations that came out of work on radar that many physicists had participated in during the war. The existence of this experimental number made clear to theorists that higher-order QED effects were real and needed to be studied. The second such higher-order effect that could be measured involved the electron's magnetic moment. In 1947, Polykarp Kusch, at Columbia, was the first to perform this experiment, with the result 1.00114 ± 0.00004 for the ratio of the actual electron magnetic moment to the prediction from the first-order QED calculation. From this one sees that the first-order QED calculation is quite good, but is off by a tenth of a percent. To the next-higher order in α, the renormalized QED calculation gives a prediction of $(1+\alpha/2\pi) = 1.00116$, in excellent agreement with Kusch's experiment. This kind of experiment has been done with higher and higher accuracy over the years,[2] with the most recent result $1.001159652189 \pm 0.000000000004$. The renormalized QED calculation has been carried out up to terms involving three powers of α and gives[3] $1.001159652201 \pm 0.000000000030$, in close agreement with the experimental value. This kind of phenomenally close agreement between an experimentally observed number and the predictions of renormalized QED is an indication of the striking success of QED as a fundamental theory of nature.

While mathematicians such as Hilbert, Weyl, and others followed closely the birth of quantum mechanics, quantum field theory was another story. Mathematically, the space of all fields is infinite-dimensional, and physicists were trying to construct operators associated with every element in this infinite-dimensional space. Very little was known to mathematicians about this kind of problem, and the techniques used by physicists were often explicitly ad hoc and self-contradictory. One particular problem was the whole subject of renormalization, which seemed to involve trying to make sense of something that was infinite by subtracting something else infinite according to somewhat unclear rules. Mathematicians found such sorts of calculations mystifying and quite alien to their own world, in which they always tried to work with rigorously well-defined concepts.

Among most physicists the attitude developed that mathematics was pretty much irrelevant to their subject. As we have seen, the mathematics of groups and their representations was often referred to as the "plague of group theory" (*Gruppenpest*) and ignored as much as possible. According to the mathematical physicist Res Jost, "In the thirties, under the demoralizing influence of quantum-theoretic perturbation theory, the mathematics required of a theoretical physicist was reduced to a rudimentary knowledge of the Latin and Greek alphabets."[4]

A detailed and excellent recent history of the development of QED, *QED and the Men Who Made It*, by physicist Silvan Schweber,[5] mentions only three mathematicians, each of them only in passing. Two of them are Weyl and Atiyah (whose names are misspelled), and the third is Harish-Chandra, who appears only in the following well-known story about Dyson from 1947:

> Harish-Chandra, who up to that time had been a physics student at the Cavendish, made the following remark: "Theoretical physics is in such a mess, I have decided to switch to pure mathematics," where-upon Dyson remarked, "That's curious, I have decided to switch to theoretical physics for precisely the same reason!"[6]

Harish-Chandra had been a student of Dirac, and went on to an illustrious career in mathematics at Columbia University and the Institute for Advanced Study. He worked mostly on representation theory, extending Weyl's work to more difficult classes of Lie groups whose interesting representations are infinite-dimensional. The two attitudes embodied by Dyson and Harish-Chandra show the barrier that quantum field theory was responsible for creating between the subjects of physics and mathematics. That the methods of quantum field theory were mathematically nonrigorous and not entirely coherent was recognized by everyone. Physicists treated this as something about which to be proud, and mathematicians as a good reason to work on something else.

Some mathematicians and mathematically minded physicists did attempt to come up with a precise and well-defined version of quantum field theory, but for a long time the only theories that could be treated in this way were the quantum field theories for free, noninteracting particles. During the 1940s, 1950s, and 1960s the subjects of physics and mathematics very much went their separate ways. This was a period of great progress in both fields, but very little contact between the two.

Further Reading

There are quite a few textbooks on quantum field theory, all aimed at graduate students in physics. The ones that have been most popular during the last 30–40 years are, in chronological order:

Bjorken and Drell, *Relativistic Quantum Mechanics*[7] and *Relativistic Quantum Fields.*[8]

Itzykson and Zuber, *Quantum Field Theory.*[9]

Ramond, *Field Theory.*[10]

Peskin and Schroeder, *An Introduction to Quantum Field Theory.*[11]

Some other suggested readings about quantum field theory:

A recent very readable and chatty introduction to quantum field theory is Zee's *Quantum Field Theory in a Nutshell.*[12]

For the history of quantum field theory and QED, see Schweber, *QED and the Men Who Made It.*

A popular book about QED is Feynman, *QED: The Strange Theory of Light and Matter.*[13]

An interesting collection of articles about quantum field theory is Cao, *Conceptual Foundations of Quantum Field Theory.*[14]

A book about the simplest form of quantum field theory and its interpretation is Teller, *An Interpretive Introduction to Quantum Field Theory.*[15]

An encyclopedic three-volume set is *The Quantum Theory of Fields I, II, III*, by Steven Weinberg.[16]

Gauge Symmetry and Gauge Theories

There are many different kinds of quantum field theories, but the ones that have turned out to be the most interesting, both for mathematics and physics, are known as gauge theories. Gauge theories have a symmetry called gauge symmetry, and again Hermann Weyl plays an important part in the story.

Einstein's general theory of relativity appeared in its final form in 1915. It postulated that four-dimensional space-time was curved and used this curvature to explain gravitational forces. One of Einstein's guiding principles in developing the theory was the principle of general covariance, which demanded that the theory be invariant under symmetry transformations corresponding to local changes in the coordinates used to parameterize space and time. If one focuses attention on a given point in space and time, nearby points are described by four coordinates. The general covariance principle is that physics should not depend on these coordinates, only on intrinsic geometrical quantities such as the distance between two points, or the curvature of space and time at a point.

In 1915 only two forces were known: gravity and electromagnetism. Quite a few physicists wanted to generalize Einstein's theory to a unified theory of gravity and electromagnetism by somehow extending the geometric framework of general relativity to include electromagnetism. Weyl noticed that one could derive Maxwell's equations for electromagnetism by extending the symmetry principle of general covariance by a new symmetry principle he called gauge invariance. Weyl's new principle of gauge symmetry allowed the use of different distance scales (or gauges) at different points. One could use a different gauge to measure sizes at different points, as long as one had a mathematical entity called a connection that related what one was doing at neighboring points. Weyl's idea was that this connection could be identified with the electromagnetic field.

The mathematical notion of a connection had also made an appearance in general relativity. In that case, a connection told one how to "parallel transport" a reference grid from one neighboring point to another. If two neighbors had separate three-dimensional grids for measuring the coordinates of objects, the connection gave the rotation necessary to relate the two grids. To do this one would have to transport one grid to the position of the other, carefully keeping the coordinate axes parallel with their position at the starting point. If one moves around on the surface of a two-dimensional space, using two-dimensional grids to measure the coordinates of points on the surface, one finds that if one parallel transports a grid around a large circle, it may or may not line back up with the original when one gets back to where one started. In general, it will line up only if the surface is flat. The amount by which the grid rotates as one goes around the circle is a measure of how curved the surface is.

Einstein quickly objected to Weyl's gauge principle by noting that it would imply that the size of a clock would change as it moved through regions containing electromagnetic fields. Two clocks that started out synchronized but then were moved through different magnetic fields would become unsynchronized. Experimental data on atoms moving in magnetic fields demonstrated no such phenomenon. Weyl's paper was published in 1918, together with a postscript by Einstein with his objections to the theory, stating ". . . it seems to me that one cannot accept the basic hypothesis of this theory, whose depth and boldness every reader must nevertheless admire."[1] Weyl ultimately abandoned his gauge theory of unified gravity and electromagnetism, but the idea that Maxwell's equations could be derived from the gauge principle remained a tantalizing one.

Schrödinger arrived in Zurich in 1921 to take up the chair in theoretical physics there. He shared Weyl's interest in general relativity and carefully studied Weyl's book on the subject, *Raum-Zeit-Materie (Space-Time-Matter)*,[2] the first edition of which had appeared in 1918. Weyl's book was very influential as one of the first texts that introduced most mathematicians and physicists to the new theory of general relativity. The third edition of the book (which appeared in 1919) included a discussion of Weyl's gauge theory of electromagnetism. In a paper published in 1922, Schrödinger noted a curious relation between Weyl's gauge theory and one of the ad hoc quantization principles of the "old" quantum theory that was a precursor to the later full theory that was to appear in 1925. This quantization principle stated that the stable circular motions of atomic particles were such that a certain mathematical quantity associated with their motion around the circle had to have integer values. Schrödinger found that this mathematical quantity was the same as the one that had appeared in Weyl's theory describing how much the size of an object would change as it went around the circle. He pointed out that the quantization condition could be understood if one reinterpreted Weyl's gauge principle to apply not to the sizes of things, but to the phase of something. If the phase of something changes as it goes around a circle by an integer multiple of 2π, it will be back in phase and match up with itself when it completes the circle.

It wasn't clear what to make of Schrödinger's observation, since at the time there was no obvious element of the theory with which one could associate a phase. Schrödinger's discovery of wave mechanics in 1925 solved this problem, since the wave function was a complex function with a phase. Schrödinger did not mention his 1922 work in his papers on wave mechanics, but some historians of science[3] claim that it may have been an important clue that pointed him in the right direction toward the discovery of wave mechanics.

Fritz London, a young physicist who was about to come to Zurich to work with Schrödinger, was the first to notice explicitly that the new quantum mechanics had Weyl's gauge symmetry when interpreted as a symmetry of phase transformations. He was so struck by this that he wrote jokingly to Schrödinger in December 1926 as follows:

Dear Professor,

I must have a serious word with you today. Are you acquainted with a certain Mr. Schrödinger, who in the year 1922 (*Zeits. fur Phys.,* 12) described a "bemerkenswerte Eigenschaft der Quantenbahnen" [remarkable property of the quantum orbits]? Are you acquainted with this man? What! You affirm that you know him very well, that you were even present when he did this work, and that you were his accomplice in it? That is absolutely unheard of . . . Will you immediately confess that, like a priest, you kept secret the truth which you held in your hands, and give notice to your contemporaries of all you know! . . . I think that it is your duty, after you have mystified the world in such a manner, now to clarify everything.[4]

London published a paper entitled "Quantum Mechanical Meaning of the Theory of Weyl" in 1927.

This gauge symmetry of U(1) phase transformations is a feature not only of quantum mechanics, but also of the Dirac equation and the full quantum field theory of electrodynamics, QED. A gauge symmetry is a symmetry of a field theory in which the symmetry transformations can be made independently at every point in space. Physicists refer to gauge symmetries as local symmetries, to be distinguished from global symmetries, where the same symmetry transformation is simultaneously performed at all points. The U(1) symmetry responsible for charge conservation is a global symmetry, since it involves the same phase transformation at every point. The gauge symmetry involves the same U(1) group of symmetries, but now with a different symmetry transformation at each point in space. Given a solution to the Dirac equation, the gauge symmetry transformation is given by multiplying it by a complex number of length one, a phase transformation, chosen independently at each point, making at the same time a compensating change in the so-called vector potential that describes the electromagnetic field.

Weyl returned to the idea of gauge symmetry in a remarkable paper published in 1929, called "Electron and Gravitation."[5] In this paper, Weyl argued that the theory of electromagnetism could actually be derived from the principle of gauge symmetry, which completely determines how a charged particle interacts with an electromagnetic field. In the same paper Weyl introduced several other extremely important ideas. He showed how to couple the Dirac equation consistently to a gravitational field, by demonstrating how to define in a curved background space-time the so-called spinor fields that are the solutions to Dirac's equation.

Weyl also introduced a new mathematical formalism for describing these spinor fields. Dirac's spinor fields had four complex components, described massive particles, and were symmetric under mirror reflection. Weyl instead rewrote the Dirac theory in terms of a pair of fields that have two complex components and are now known as Weyl spinors. Using just one element of the pair, one gets a theory of massless spin-one-half particles that is asymmetric under mirror reflection and was much later (in 1957) found to be just the thing to describe the neutrino and its weak interactions. The mathematical setup first worked out in this paper of Dirac spinors on a curved manifold, broken up into a pair of Weyl spinors, was to become an important part of the interaction between mathematics and physics of the late 1970s that will be discussed in a later chapter.

From the point of view of representation theory, Weyl spinors are the fundamental representations that occur when one studies the representations of rotations in four-dimensional space-time. Recall that in three dimensions the group of rotations is called SO(3), and that spin-one-half particles are representations not of this group, but of a doubled version of it, which turns out to be the group SU(2) of transformations on two complex variables. Three-dimensional geometry thus has a subtle and nonobvious aspect, since to really understand it one must study not just the obvious three-dimensional vectors, but also pairs of complex numbers. These pairs of complex numbers, or spinors, are in some sense more fundamental than vectors. One can construct vectors out of them, but one cannot construct spinors just using vectors.

When one generalizes to rotations in four dimensions, one finds that such rotations can in some sense be given by a pair of two independent three-dimensional rotations. Thus spinors in four dimensions can be built out of two different three-dimensional spinors, which explains the two types of Weyl spinors that make up the pair of spinors used by Dirac. Working together with the algebraist Richard Brauer at the Institute of Advanced Study in Princeton during 1934–1935, Weyl generalized the theory of spinors to not just three or four, but an arbitrary number of dimensions. These general spinor representations had been first discovered by the French geometer Elie Cartan in 1913. The new construction of them given by Brauer and Weyl used Clifford algebras and was inspired by Dirac's use of these algebras in the Dirac equation.

The gauge symmetry of QED is one of its physically most important and mathematically most attractive features. If handled properly, it is an extremely powerful principle. It very tightly constrains the theory, since it keeps one from having to worry about all sorts of extra terms that might need to go into the equations of the theory. Only terms that respect the gauge symmetry need to be considered. At the same time, it leads to major technical difficulties. All sorts of approximations and calculational techniques that one might like to try are unusable because they violate the gauge symmetry principle. This was one of the main reasons it took from the early 1930s to the late 1940s to work out the details of how to renormalize QED.

During the 1930s and 1940s, as physicists began studying the so-called strong forces between protons and neutrons, they came to realize that there was a familiar group of symmetries at work in this situation. If one puts the fields for the proton and the neutron together as one field described by two complex values at each point, it turns out that the strong interactions are invariant under the same group SU(2) that had appeared in the case of spin. Since this phenomenon was first observed in nuclear physics as a relation between different nuclear isotopes, it was dubbed isospin. The strong interactions between nucleons (protons and neutrons) are said to have an SU(2) isospin symmetry. These symmetry transformations don't have anything to do with spatial rotations; they are what is called a purely internal symmetry. They mix together the two complex numbers corresponding to the "proton-ness" and the "neutron-ness" of a nucleon. The strong force knows only that it is acting on a nucleon, and is the same no matter which kind (proton or neutron) it is.

In 1954, Chen Ning Yang and Robert Mills published a paper about a possible generalization of QED that they thought might be useful as a quantum field theory that could describe the strong interactions. Whereas QED has a local gauge symmetry of U(1) phase transformations, their theory had a local gauge symmetry of SU(2) isospin symmetry transformations. This generalized kind of gauge symmetry has become known as a Yang–Mills gauge symmetry, and the theory they wrote down is now called an SU(2) Yang–Mills quantum field theory. They found this generalization of the gauge symmetry principle to be very attractive, but it led to a quantum field theory with puzzling properties. In the Yang–Mills theory there was a Yang–Mills field analogous to the electromagnetic field, but its quanta came in three different kinds. So there were three different kinds of particles analogous to the photon of QED, and in addition, these three different particles had forces between them. In QED, if one forgets about the electrons and looks just at the photons, things are very simple, since the photons do not interact with each other. In Yang–Mills theory the three analogues of the photon interact with each other nontrivially in a way that is fixed by the gauge symmetry.

If one goes ahead and tries to construct the perturbation series for Yang–Mills theory along the lines of what worked for QED, one finds that to zeroth order one has a theory of nucleons and three different kinds of massless photons. Since no such triplet of massless photons had been observed, interest in the theory was very limited. Pauli had investigated the theory even before Yang and Mills, but had stopped working on it because it seemed to predict unobserved massless particles. A further problem was that the renormalization techniques that had allowed higher-order calculations in QED did not work for Yang–Mills theory. So Yang and Mills seemed to have produced a theory that did not correspond to any known physics, and was probably inconsistent to boot. They stopped working on the theory and went on to other things. Despite the problems, the Yang–Mills theory was in many respects a very attractive way to generalize QED, and a number of theorists worked intermittently during the coming years on the problem of how properly to carry out higher-order calculations in it.

Yang was at the Institute for Advanced Study at the time (1953–1954) that he was working on the Yang–Mills theory. Hermann Weyl would undoubtedly have been very interested in this generalization of his original gauge theory, but he had retired from his professorship at the Institute in 1952 and was to die in Zurich in 1955. Many of the best of a younger generation of mathematicians passed through the institute during these years, including the topologist Raoul Bott and geometer Michael Atiyah. Bott[6] recalls socializing with Yang at the time, including spending Saturday mornings with him painting the fence of the nursery school. It didn't occur to either of them that they might have anything in common in their research work. Only in the late 1970s did they both realize that a quarter century earlier they had been pursuing closely related topics without knowing it. Atiyah and Bott were to make good use of the Yang–Mills equations to solve problems in geometry and topology. Yang was to find that the mathematical language of connections and curvature that his mathematical colleagues at the institute had been exploring during the 1950s captured precisely the ideas about gauge theory that he had been working on with Mills.

A Hungarian emigré, Bott studied electrical engineering at McGill University in Montreal during the war, then later changed fields to study mathematics, enrolling as a graduate student at the Carnegie Institute of Technology in Pittsburgh. There he met Weyl, who was visiting to give a colloquium talk, and this led to an invitation to come to work at the institute in Princeton. During his time at the institute, Bott was to work out an extension of Weyl's work on representation theory, one that brought together topology, geometry, and representation theory in a very beautiful way.

The 1950s were a golden age of mathematics, especially in Princeton and in Paris. Those years saw the development of a large number of fundamental ideas about modern mathematics that remain central to this day. It was also a time of minimal contact between physicists and mathematicians, with each of the two groups discovering things whose significance would become clear to the other only many years later.

FURTHER READING

An excellent book on the history of gauge theory is O'Raifeartaigh's *The Dawning of Gauge Theory*.[7]

6

The Standard Model

The history of theoretical and experimental particle physics during the third quarter of the twentieth century is a complex story. We are just beginning to be distant enough from it in time to be able to bring it into perspective. Telling this story in any detail would require a separate book, and several good ones on the subject already exist. Two that are particularly recommended are Crease and Mann's *The Second Creation*[1] and Riordan's *The Hunting of the Quark,*[2] both of which do a wonderful job of making this history accessible.

In retrospect, it is clear that far and away the most important aspect of this history was the formulation of something that came to be known as the standard model. The essence of this model for particle physics was in place by 1973, and at the time it was considered the simplest of a class of many possible particle theory models. By the end of the 1970s, it had been confirmed by many experimental results and was without serious competition. The designation "standard model" first starts to appear in the titles of scientific papers in 1979 and was in widespread use for a few years before that.

What is this "standard model"? To answer this question fully requires going over the content of what is by now a course usually taught to physics graduate students in the second year or so of a doctoral program, but an attempt will be made in this chapter at least to indicate some of the main ideas. In simplified form, the standard model starts with QED, which is a U(1) gauge theory, and extends it by two new Yang–Mills gauge theories. One of these uses the group SU(2) and describes weak interactions; the other uses SU(3) and describes strong interactions. Besides the three kinds of forces coming from these three gauge theories, the other main component of the theory is the specification of the particles that experience these forces. These particles come in three nearly identical sets called generations, where the only difference between the corresponding particles in two different generations is their mass. The lowest-mass genera-tion contains all the stable particles that make up our everyday world; they are the electron, the electron neutrino, and the up and down quarks.

One aspect of the standard model still remains mysterious. This is the fact that the vacuum is not symmetric under the SU(2) symmetry group of the weak interaction gauge theory. A separate section will examine this phenomenon in more detail.

THE STANDARD MODEL: ELECTROWEAK INTERACTIONS

From the time of its first formulation in 1927–1929, QED was a very promising explanation of atomic physics and the electromagnetic force, but there were two other sorts of forces known that seemed to have nothing at all to do with electromagnetism or QED. One of these is the strong force that binds protons together in the nucleus. It is stronger than electromagnetism, overcoming the fact that protons are all positively charged and thus would fly apart were it not for the existence of a stronger force that holds them together.

There is also a much weaker force than electromagnetism, one that is responsible for the radioactive decay of certain nuclei, first discovered by Henri Becquerel in 1896. The proper understanding of this so-called beta decay, in which an electron is observed to be emitted from a nucleus, began with Pauli's suggestion in 1930 that the energy spectrum of these decays could be explained if an unob-served light, uncharged particle called a neutrino was being emit-ted at the same time as the electron. The earliest theory of this weak interaction was due to Enrico Fermi in 1933, and a later ver-sion was completed in 1957 simultaneously and independently by several physicists. It became known as V-A theory, a reference to the terms vector and axial vector, which describe the symmetry properties of the interaction under the symmetries of rotation and mirror reflection.

V-A theory had the peculiar property of being chiral, i.e., having a handedness. This means that the theory is not symmetric under the mirror-reversal symmetry discussed earlier as the simplest example of a group representation. If one looks at one's right hand in a mirror it appears to be a left hand and vice versa. Physicists had always assumed that the sort of reversal that occurs when one looks at things in a mirror had to be a symmetry transformation of any fundamental physical law. It turns out that this is not the case for the weak interactions, a fact that remains somewhat mysterious to this day. V-A theory says that the natural fields that occur in the theory of the weak interactions are not Dirac's four-component spinors, but Weyl's two-component ones.

Even though V-A theory was quite successful at explaining the observed properties of the weak interaction, it had the same kind of infinities that had first discouraged physicists about quantum field theory. While the first-order terms in the perturbation series gave good results, higher-order terms were all infinite and could not be renormalized in the way that had been done in QED. Since the interaction strength was so weak, these higher-order effects were likely to be too small to be measured, so the inability to calculate them was not of any practical significance. This nonrenormalizability remained a significant concern, putting the consistency of the theory in doubt.

In QED, the electromagnetic force is carried by the photon, and two charged particles experience a force because they both interact with the electromagnetic field, whose quanta are the photons. The weak interactions seemed to be quite different, since there was no field analogous to the electromagnetic field to propagate the interaction from place to place. In V-A theory, weak interactions took place between particles coming together at the same point. It was known, however, that one could make V-A theory much more like QED, but to do so one needed the field that transmitted the weak force to be of very short range. The quanta of this field would be very massive particles, not massless ones like the photon. If one conjectured that these quanta had a mass on the order of 100 GeV, then the actual strength of interaction of particles with the field could be of the same strength as the analogous electromagnetic interaction. But accelerators of the day were barely able to produce particles with masses of hundreds of MeV, so the conjectural quanta of the weak interactions with a thousand times greater mass were far out of reach.

The idea of constructing a theory in which the weak and electro-magnetic interactions were of the same strength and thus could be unified was pursued by Julian Schwinger, who wrote about it in 1957 and lectured to his students at Harvard on the topic during the mid-1950s. One of these students, Sheldon Glashow, took up the idea and in 1960 came up with a precise model. The kind of theory Glashow proposed did not have gauge symmetry, but it was essentially the Yang–Mills theory, with an added term to give mass to the quanta of the weak field. This added term ruined the gauge symmetry. The theory also continued to have problems with infinities and could not be renormalized by any known methods. The symmetry group of Glashow's theory had two factors: one was the original U(1) phase transformation symmetry of QED; the other was the SU(2) of Yang and Mills. This symmetry group with two factors is written SU(2) × U(1).

One more idea was needed to fix the gauge symmetry problems of the Glashow model and make it consistent. What was needed is something that has come to be known as a Higgs field, and in the next section this idea will be examined in detail. In the fall of 1967, Steven Weinberg came up with what was essentially Glashow's model for a unified model of the weak and electromagnetic interactions, but aug-mented with a Higgs field so that the quanta carrying the weak force would be massive while maintaining the gauge symmetry of the the-ory. The same idea was independently found by Abdus Salam, and this kind of unified model of electroweak interactions is now known variously as the Weinberg–Salam or Glashow–Weinberg–Salam model.

Weinberg and Salam conjectured that this theory could be renor-malized, but neither was able to show this. Since the initial sugges-tion of Yang and Mills, several people had tried to construct the perturbation expansion for the theory and renormalize it, but this turned out to be difficult. The techniques developed to deal with gauge invariance in QED could not handle the generalized gauge invariance of the Yang–Mills theory. Finally, in 1971, Gerard 't Hooft, working with his thesis advisor Martin Veltman, was able to show that Yang–Mills theories are renormalizable. In particular, the Glashow–Weinberg–Salam model SU(2) × U(1) Yang–Mills theory is a renormalizable quantum field theory, and in principle one could consistently calculate terms of any order in its perturbation expansion.

PHYSICS DIGRESSION: SPONTANEOUS SYMMETRY BREAKING

In discussing the importance of symmetry in physics, the symmetry transformations considered so far have been ones that leave the laws of physics invariant. Here the "laws of physics" means the dynamical laws that govern how the state of the world evolves in time, expressed in classical physics by Newton's laws and Maxwell's equations, and in quantum physics by the Schrödinger equation. A subtle point about this is that while the form of the equations may not change under symmetry transformations, in general the solutions to the equations will change. While the laws governing the evolution of the state of the world may be symmetric, the actual state of the world generally is not. If one looks around, things are not rotationally symmetric, even if the laws of physics are.

While this lack of symmetry may apply to randomly chosen states, physicists generally assumed that the vacuum state was always symmetric. No matter what symmetry transformations one performed, they would not change the vacuum state. This attitude started to change during the 1950s, largely under the influence of ideas from a field seemingly far removed from particle physics, that of condensed matter physics. This field also goes by the name of solid-state physics, and one of its main topics is the study of the effects of quantum mechanics on the physics of atoms and electrons in solids. Unlike particle physicists, solid-state physicists are interested not in elementary particles and the interactions between them, but in the behavior of very large numbers of particles, interacting together to form a macroscopic amount of matter. Particle physicist Murray Gell-Mann famously and dismissively referred to the field as "squalid-state" physics, since its objects of study are inherently much more complex and messier than elementary particles.

By the 1950s, condensed matter physicists had found that quantum field theory was a useful tool not just for studying elementary particle physics, but also for the very different kinds of problems that concerned them. One example of where quantum field theory is helpful is in the problem of understanding the thermodynamics of a solid at very low temperature. Atoms in a solid are often bound into a regular lattice, but can move slightly in vibrating patterns of various kinds. The quantization of this vibrational motion leads to new quanta called phonons, which formally are roughly analogous to photons. Whereas photons are meant to be quanta of a fundamental field, phonons don't come from a truly fundamental field, but from the motion of the atoms of the lattice.

In a quantum field theory treatment of a condensed matter problem, the analogue of the vacuum is not the state with no particles, but instead, whatever the lowest-energy state is for the large number of atoms one is considering. A major goal of condensed matter physics is to understand what this state is and how to characterize it. Unlike the case of the vacuum in particle theory, here there are obvious reasons why the lowest-energy state may not be invariant under symmetry transformations of the theory. A standard simple example of symmetry breaking in the lowest-energy state involves ferromagnetic materials. These are solids in which each atom has a magnetic moment. In these materials one can think of each atom as having a vector attached to it that points in some direction. The lowest-energy state will be that in which all the vectors line up in the same direction. There is no preferred direction in the problem, since the laws that govern the dynamics of how these vectors interact are symmetric under rotational transformations, but even so, the lowest-energy state is not rotationally symmetric.

This kind of phenomenon, where the lowest-energy state is not invariant under the symmetries of the theory, has acquired the name "spontaneous symmetry breaking," and a theory in which this happens is said to have a spontaneously broken symmetry. During the 1950s, various particle theorists considered the possibility that this same idea could be applied in particle physics. The kind of application they had in mind was to try to explain why certain symmetries, in particular the SU(2) isospin symmetry, seemed to hold in nature, but only approximately. Perhaps the vacuum state was not invariant under the isospin symmetry, so there would be spontaneous symmetry breaking and this would explain what was going on.

What particle theorists soon discovered was that as a general rule, the effect of spontaneous symmetry breaking was not to turn exact symmetries into approximate ones. Instead, spontaneous symmetry breaking was characterized by the existence in the theory of massless spin-zero particles. These are now known as Nambu–Goldstone particles, after the physicists Yoichiro Nambu and Jeffrey Goldstone. They come about because one can use the existence of an infinitesimal symmetry transformation that changes a vacuum state to define a new field, one with the characteristic that its quantum field theory must have zero-mass particles. Goldstone gave this argument in 1961, and the necessity of zero-mass particles in the case of spontaneous symmetry breaking became known as Goldstone's theorem.

One of the great successes of quantum-field-theoretical methods in condensed matter physics was the successful BCS theory of superconductivity developed by John Bardeen, Leon Cooper, and Robert Schrieffer in 1957. In a superconductor there is an indirect force between electrons due to the fact that each electron will distort the positions of atoms in the solid, which in turn will have an effect on other electrons. This force between the electrons can be such that they are able to lower their energy by pairing up and acting in a coherent way. When this happens, the lowest-energy state of the material is quite different from what is expected. Instead of having stationary electrons, it has pairs of electrons moving in a synchronized fashion, and currents can flow through the material with no resistance. This surprising ability to carry electric current with no resistance is what characterizes a superconductor.

It was realized early on that the BCS superconductivity theory was a theory with spontaneous symmetry breaking. The lowest energy state was very non-trivial, containing pairs of electrons that behaved coherently. This kind of spontaneous symmetry breaking was called dynamical spontaneous symmetry breaking, since its origin was in a dynamical effect (the indirect force between electrons) that changed dramatically the lowest-energy state from what one would naively expect. An obvious question to ask was, what symmetry is being broken by this new vacuum state? The answer turned out to be just the well-known U(1) gauge symmetry of electrodynamics. To describe what was going on, a field whose quanta were the coherent electron pairs was introduced, with the U(1) gauge symmetry transformations acting on this field. The dynamics of the theory were invariant under the U(1) gauge transformations, but the vacuum state was not.

The whole issue of how to treat the gauge symmetry of a superconductor was initially very confusing, but by 1963 the condensed matter theorist Philip Anderson had shown how to do this. In the process he explained another unusual aspect of superconductors called the Meissner effect. This is a property of superconductors that causes them actively to exclude magnetic fields from their interior. If one puts a superconductor in a magnetic field, the field falls off exponentially with a characteristic length as one goes inside the material. This is the behavior one would expect if the photon were massive. What Anderson showed was that, as far as electromagnetic fields were concerned, a superconductor could be thought of as a different kind of vacuum state in which they could propagate. In this new vacuum state the U(1) gauge symmetry of electromagnetism was spontaneously broken, and the effect of this was not the zero-mass particles predicted by the Goldstone theorem. Instead, the photon became massive. In some sense, the massless photon combines with the massless Nambu–Goldstone particles to behave like a massive photon. In his 1963 paper, Anderson also noted that the same mechanism might work in Yang–Mills theory with its more general SU(2) gauge invariance.

Several physicists took up this suggestion and tried to see whether they could make elementary particle theories with Yang–Mills gauge symmetries that were spontaneously broken, giving a mass to the analogues of the photon in Yang–Mills theory. One of these physicists was a Scotsman, Peter Higgs, who in 1965 wrote down such a theory (the Belgian physicists Robert Brout and Francois Englert did similar work around the same time). To do this, Higgs had to introduce a new field, the analogue of the field describing coherent pairs of electrons in the superconductor. This field is introduced in such a way that it will automatically cause spontaneous symmetry breaking of the gauge symmetry of the theory and is now referred to as a Higgs field. Note that this kind of spontaneous symmetry breaking is not really dynamical. It is not an indirect effect caused by the dynamics of the theory, but comes from choosing to introduce a new field with exactly the right properties. This way of getting around the Goldstone theorem and giving mass to the Yang–Mills field quanta is now called the Higgs mechanism, and it was this idea that Weinberg and Salam used in 1967 to turn Glashow's earlier model into the one that would be the basis of the electroweak part of the standard model.

The Standard Model: Strong Interactions

During the 1950s and 1960s, while progress was being made in using quantum field theory to understand the weak and electromagnetic interactions, the strong interaction that binds nucleons into the nucleus seemed to be a completely different story. As ever-higher-energy particle accelerators were constructed, allowing ever-higher-energy nucleon-nucleon collisions to be studied, a bewildering array of new particles was being produced. In a quantum field theory such as QED, studied in a perturbation expansion, the different kinds of particles one sees correspond in a one-to-one fashion with the fields of the theory. A quantum field theory of the strong interactions would appear to require an ever increasing number of fields, and there seemed to be no obvious pattern relating them.

The only symmetry that the strong interactions were known to have was the isospin SU(2) symmetry, the one that motivated Yang and Mills. Attempts were made to generalize this symmetry to try to put some order into the many particle states that had been discovered, with the first success of this kind being due to Yuval Ne'eman and Murray Gell-Mann in 1961. They discovered that the strongly interacting particles fit into representations not just of SU(2), but of the larger group SU(3), the group of special unitary transformations of three complex numbers.

The representation theory of SU(3) was an example of the general theory of representations of Lie groups brought to completion by Weyl in 1925–1926, and this was by 1960 a rather well known subject among mathematicians. During the academic year 1959–1960, Gell-Mann was at the Collège de France, trying to generalize the SU(2) isospin symmetry, and he often had lunch with the local French mathematicians.[3] At least one of them, Jean-Pierre Serre, was one of the world's experts on the type of representation theory that Gell-Mann needed. It never occurred to Gell-Mann to ask his lunchtime companions about the mathematical problems he was encountering. The 1950s and 1960s were very much a low point in the history of encounters between the subjects of mathematics and physics. Most mathematicians were working on problems that seemed to them far from physics, and the opinion among physicists was pretty much that whatever the mathematicians were up to, it couldn't be anything very interesting or of any possible use to them.

Late in 1960, back at Caltech, Gell-Mann did finally talk to a mathematician (Richard Block) and found out that what he was trying to do was a well-understood problem in mathematics that had been solved long ago. Once he knew about SU(3) and what its representations were, he was able to show in 1961 that the observed particles fit into patterns that corresponded to some of these representations. The simplest representation of SU(3) that occurred was an eight-dimensional one, so Gell-Mann started calling SU(3) symmetry the "Eightfold Way" in a joking allusion to the eightfold path of Buddhism. Among the many particles that did not fit into eight-dimensional representations, Gell-Mann identified nine of them that could be fit into a ten-dimensional representation, with one missing. He was able to use representation theory to predict the properties of a hypothetical tenth particle that would be the missing particle in the representation. In 1964, this particle, the omega-minus, was discovered with the predicted properties in the 80-inch bubble chamber at Brookhaven.

From this time on, some exposure to Lie groups and their representations became part of the standard curriculum for any particle theorist. I first learned about the subject in a graduate course at Harvard taught very ably and clearly by Howard Georgi. While physicists were by then well aware of how useful representation theory could be and were no longer talking about the *Gruppenpest*, there was still a strong feeling of wariness toward the whole subject. In the book that Georgi later wrote based on his lecture notes from this course, he warns,

A symmetry principle should not be an end in itself. Sometimes the physics of a problem is so complicated that symmetry arguments are the only practical means of extracting information about the system. Then, by all means use them. But, do not stop looking for an explicit dynamical scheme that makes more detailed calculation possible. Symmetry is a tool that should be used to determine the underlying dynamics, which must in turn explain the success (or failure) of the symmetry arguments. Group theory is a useful technique, but it is no substitute for physics.[4]

His introduction to the book is even more explicit:

I think that group theory, perhaps because it seems to give information for free, has been more misused in contemporary particle physics than any other branch of mathematics, except geometry. Students should learn the difference between physics and mathematics from the start.[5]

One remaining mystery was that the representations being used to classify strongly interacting particles did not include the obvious one that is involved in the definition of the SU(3) symmetry group: the fundamental representation on a triplet of complex numbers. All representations of SU(3) can be built out of these triplets, but if they corresponded to particles they would have to have electric charges that were not integers but instead fractions, multiples of one-third of the electric charge of a proton or electron. Such fractionally charged particles had never been seen. Gell-Mann finally came around to the idea that the possibility of the existence of such particles should be taken seriously and gave them a name: quarks. He identified the source of this made-up word as a line in James Joyce's *Finnegans Wake*, which reads, "Three quarks for Muster Mark!" Another physicist, George Zweig, came up with a similar proposal, in his case calling the new fractionally charged particles "aces."

After the successes of SU(3) representation theory in classifying particles, physicists quickly started making up for lost time and began looking for more general symmetry groups. Attempts were made to find symmetry groups that brought together the internal SU(3) symmetry group with the SU(2) group of rotations in space. One example that was studied was the symmetry group SU(6), which is large enough to contain both SU(3) and SU(2) symmetry transformations. By 1967 this work mostly came to a halt after Sidney Coleman and Jeffrey Mandula gave an argument now known as the Coleman–Mandula theorem. Their argument showed that any theory that combined the rotational symmetry group and an internal symmetry group into a larger symmetry group such as SU(6) would have to be a trivial theory in which particles could not interact.

During the mid-1960s, Gell-Mann's SU(3) symmetry was further developed and extended, using a set of ideas that, for reasons that would require an extensive digression to explain, were given the name "current algebra." One aspect of the current-algebra idea was that the strong interactions were conjectured to have not one, but two different SU(3) symmetries. The first SU(3) is the same whether or not one does a mirror-reversal transformation. This is Gell-Mann's original SU(3), and the vacuum state is left unchanged by these SU(3) symmetry transformations. This is the symmetry that leads to the classification of particles according to SU(3) representations. The second SU(3) is one that in some sense changes sign when one does a mirror-reversal. The vacuum is not invariant under this SU(3). It is a spontaneously broken symmetry. The spontaneous breaking is presumed to be a dynamical effect of the strong interactions, so this is a case of dynamical spontaneous symmetry breaking. Since this is an example of spontaneous symmetry breaking not involving a gauge symmetry, Goldstone's theorem applies. As a result, there should be eight massless Nambu–Goldstone particles, corresponding to the eight dimensions of SU(3). Eight pions have been found that have the right properties to be the Nambu–Goldstone particles for this symmetry, but they are not massless. On the other hand, these eight particles are of much lower mass than all the other strongly interacting particles, so this picture is at least approximately correct.

Various calculational methods were developed that allowed physicists to use current algebra to calculate properties of pions and their interactions. In general, these gave approximately correct predictions, although a couple of the predictions of the theory were badly off. Further study of these cases revealed some subtleties, and the failure of the naive symmetry arguments in certain situations was called the chiral anomaly. Here the term "chiral" refers to the fact that the symmetry in question changes sign under mirror-reversal. The term "anomaly" has come to refer to the phenomenon of the failure of standard arguments about symmetries in certain quantum field theories. This phenomenon was to be investigated by mathematicians and physicists in much greater detail over the next two decades.

Starting in 1967, experimentalists using the new 20-GeV linear electron accelerator at SLAC began a series of experiments designed to measure the scattering of electrons off of a proton target. The great majority of the collisions of this kind lead to a relatively small transfer of momentum between the electron and the target, and it was these that had been the focus of most experiments. The SLAC experimentalists were able instead to measure scatterings that involved a large momentum transfer. These were known as "deep inelastic" scatterings. Here "deep" refers to the large momentum being transferred from the electron to the target, "inelastic" to the fact that the transfer of momentum to the proton leads to particle production, not just one particle bouncing off another.

Over the next few years, the SLAC experiments gave very unexpected results, with from ten to a hundred times more particles scattering at large momentum transfer than expected. One interpretation of what they were seeing was that the proton was not a structureless object of size about 10^{-15}m, as expected, but instead had pointlike constituents. This was very much analogous to what had happened sixty years earlier when Ernest Rutherford had discovered the atomic nucleus by scattering alpha-particles off an atom and seeing an unexpected amount of scattering at large angles.

In the case of the nucleus, it was clear what was going on, since with enough energy one could actually knock the nucleus out of an atom and study it separately. In the SLAC experiments there was no sign of any new pointlike particles among the products of the collisions. This situation became more and more confusing as better data from the experiment showed that deep inelastic scattering data were showing the property of scaling. This meant that not only did there seem to be pointlike constituents, but these constituents of the proton were behaving like free particles, interacting only weakly. One would have liked to conjecture that these constituents were the quarks, but it was well known that there had to be very strong forces binding the quarks together in the proton, forces so strong that a single quark could not be removed and observed separately.

By late 1972, David Gross and Sidney Coleman had embarked on a project of trying to prove that the SLAC results could not be interpreted within the framework of quantum field theory. They were inspired by a new interpretation of the whole issue of renormalization in quantum field theory due to Kenneth Wilson, who had noticed a close analogy between renormalization and the theory of phase transitions in condensed matter physics. Wilson's idea was that the interaction strength in quantum field theory should be thought of not as a single number, but as something that depends on the distance scale being probed. In Wilson's picture, a quantum field theory at each distance scale has an "effective" interaction strength, in which the effects of smaller distances are in some sense averaged out. If one knows the effective interaction strength at a fixed distance scale, one can calculate it at larger distance scales by averaging over the effects occurring at sizes in between the fixed and larger scales. From this point of view, the older renormalization calculations have to do with the relationship between the bare interaction strength, which corresponds to taking the distance scale to zero, and the physical interaction strength, which corresponds to the large distance scales at which most experiments are performed.

In QED, the large-distance interaction strength is given by the parameter $\alpha = \frac{1}{137}$, which also governs the relative size of terms in the perturbation expansion. If one tries to determine what the interaction strength at shorter distances should be for QED, one finds that the effective interaction strength grows at shorter distances. At very small distances the perturbation expansion will become useless once the interaction strength gets large enough that succeeding terms in the expansion no longer get smaller and smaller. This problem had been known since the 1950s, and had often been taken as indicating that for short enough distances, quantum field theory would have to be replaced by something else. Gross set out to show that all quantum field theories behaved like QED, so that none of them could ever hope to explain the SLAC observations of an interaction strength that was getting small at short distances. While most quantum field theories could be shown to behave like QED, there was one theory in which the calculation was quite tricky. This theory was the Yang–Mills theory that 't Hooft and Veltman had recently shown was renormalizable and thus, in principle, amenable to calculation.

Gross was by this time teaching at Princeton, where Coleman joined him for the spring semester of 1973 while on leave from Harvard. Frank Wilczek had arrived in 1970 at Princeton as a mathematics graduate student, but transferred to the physics department after being inspired by taking Gross's quantum field theory course. Wilczek became Gross's first graduate student, and Gross was later to comment, "He spoiled me. I thought they'd all be that good."[6] Wilczek went to work on the problem of calculating the behavior of the Yang–Mills effective interaction strength, with Gross assuming that he would find the same behavior as in all other quantum field theories. Gross was expecting Wilczek not to discover something new, but just to finish off the proof that quantum field theory could not explain the SLAC results.

That spring, a student of Coleman's at Harvard named David Politzer also took up the same calculation. When he completed it, he found that the behavior of Yang–Mills theory was opposite to that of all other quantum field theories. He called Coleman at Princeton to tell him about this, and Coleman informed him that Gross and Wilczek had just completed the same calculation, but their result was that the Yang–Mills behavior was the same as in other theories. Politzer checked his calculation and could find no error, and around this time Gross and Wilczek located a sign error in their work.

The final result of both calculations was that the effective interaction strength in Yang–Mills theory becomes smaller, not larger, as one goes to shorter and shorter distances. This new behavior was given the name "asymptotic freedom": at shorter and shorter (asymptotically small) distances, particles behave more and more as if they were moving freely and not interacting with each other. The significance of the result was immediately clear. A Yang–Mills theory could have an effective interaction strength that was strong at long distances, binding the quarks together, and weak at short distances, as required by the SLAC experiments. The flip side of asymptotic freedom at short distances was the fact that the interaction strength became larger and larger at longer distances. This mechanism, in which a force grows larger at large distances, thus keeping quarks permanently bound together, is sometimes called "infrared slavery."

A quantum field theory of the strong interactions was finally at hand. It had been known for a while that to get Gell-Mann's quark model to agree with experiment, each of Gell-Mann's original three quarks had to come in three varieties, and the property that differentiated these three varieties was given the name "color" (the two "threes" here have nothing to do with each other). Turning the symmetry among these three colors into a gauge symmetry gave an SU(3) Yang–Mills theory, and this theory was quickly given the name quantum chromodynamics and the acronym QCD to honor its close relationship as a quantum field theory to QED.

The QCD SU(3) symmetry is a gauge symmetry and is not related in any way to the two SU(3) symmetries first studied by Gell-Mann and used in current-algebra calculations. Those latter two SU(3) symmetries are global symmetries, not gauge symmetries, and in any case are only approximate. The property that distinguishes Gell-Mann's original three kinds of quark from one another is now called "flavor" (a randomly chosen attribute different from color). The three flavors of quark known to Gell-Mann are called "up," "down," and "strange" (that these names are not names of flavors as the word is usually understood is one of many inconsistencies of the terminology). The current-algebra SU(3) groups are symmetry transformations that transform quarks of one flavor into quarks of another flavor, not changing the colors at all. If all of the masses of the quarks were the same, then one of the current-algebra SU(3) groups (the mirror-reversal symmetric one) would be an exact symmetry. If all of the masses of the quarks were zero, both of the current-algebra SU(3) groups would be exact symmetries. In retrospect, the approximate success of current-algebra predictions was due to the fact that while the three flavors of quark all had nonzero masses, these masses were much smaller than the overall mass scale of QCD, which corresponds to the distance scale at which the QCD interactions become strong. This whole picture lent added credence to QCD, since it provided not only a plausible explanation of why there were current-algebra SU(3) symmetries, but also why they were only approximately valid.

QCD has a remarkable property shared by no other physical theory. If one ignores the masses of the quarks, there is only one free parameter, the one that governs the interaction strength. But the lesson of Wilson's approach to renormalization is that this is not really a parameter. It depends on a choice of distance scale, and if one sets it to some number at a fixed distance scale, one can calculate what it will be at all others. So, in QCD one's choice of distance units and one's choice of parameter are linked, with one determining the other. This behavior was christened "dimensional transmutation" by Coleman (who also gave an alternative terminology: "dimensional transvestitism"). QCD with no quark masses is thus a completely uniquely determined theory, and there are no parameters in the theory that can be adjusted. This sort of uniqueness is a goal that theoretical physicists always dream of achieving. Ideally, one wants a theory to be able to predict everything, without one's having to choose some parameters in the theory to make things agree with experiment. QCD gets closer to this than any other known theory, and this was one reason it quickly became very popular.

FURTHER READING

Besides *The Hunting of the Quark*,[7] by Riordan, and *The Second Creation*,[8] by Crease and Mann, for a more technical point of view there are two excellent collections of articles, mostly written by the physicists directly involved in the work described:

Pions to Quarks: particle physics in the 1950s.[9]

The Rise of the Standard Model: Particle Physics in the 1960s and 1970s.[10]

For a recent review of the history of the standard model by one of the participants in this story, see "The Making of the Standard Model,"[11] by Steven Weinberg. This article is contained in a volume entitled *50 Years of Yang–Mills Theory*,[12] edited by Gerard 't Hooft, which has quite a few informative review articles about aspects of the standard model, together with his commentary.

Triumph of the Standard Model

With the advent of asymptotic freedom and QCD in the spring of 1973, the set of ideas needed for the standard model was complete. The strong interactions were to be described by quarks interacting via QCD (an SU(3) Yang–Mills theory), and the weak interactions were to come from the Glashow–Weinberg–Salam model (an SU(2) × U(1) Yang–Mills theory). There remained one major problem: for the Glashow–Weinberg–Salam model to be consistent with what was known about quarks, Glashow (together with John Iliopoulos and Luciano Maiani) had shown in 1970 that one needed to add a fourth flavor of quark to the three known to Gell-Mann. This conjectural fourth flavor of quark was called the "charmed" quark, but there was no experimental evidence for it. It would have to be much more massive than the other quarks to have escaped the notice of experimenters.

Altogether, the standard model now had four flavors of quarks (up, down, strange, and charmed), each of which came in three colors, and four leptons. The leptons were particles that have no strong interactions: the electron and electron neutrino, and the muon and muon neutrino. These particles could be naturally organized into two generations. The first generation consists of the up and down quarks and the electron and electron neutrino. These particles are those of lowest mass and are all that is needed to make up protons, neutrons, and atoms, essentially all of the everyday physical world. The second generation of particles (strange and charmed quarks, muon, and muon neutrino) is in every way identical to the first generation except that the particles have higher masses. As a result they are unstable, and if produced will soon decay into the lower-mass particles of the first generation.

The first major success of the standard model came with the unexpected experimental discovery in November 1974 of a particle

Leptons	Quarks

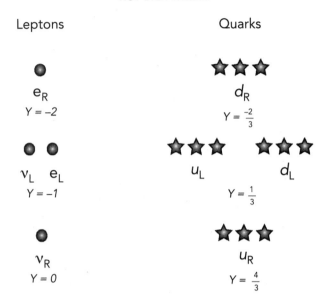

The First Generation of Standard Model Fermions. This picture shows the SU(3) × SU(2) × U(1) transformation properties of the first of three generations of fermions in the standard model. (The other two generations behave the same way.)

Under SU(3), the quarks are triplets and the leptons are invariant.

Under SU(2), the particles in the middle row are doublets (and are left-handed Weyl spinors under Lorentz transformations), and the other particles are invariant (and are right-handed Weyl spinors under Lorentz transformations).

Under U(1), the transformation property of each particle is given by its "weak hyperchange" Y.

called the J/psi. This event and what followed from it came to be known as the November Revolution. Experimentalists at Brookhaven and SLAC, using very different techniques, discovered nearly simultaneously a particle that the Brookhaven group named the J, the SLAC group the ψ (or psi). The discovery was particularly striking at SLAC, where the particle appeared as a resonance, produced copiously at a very specific beam energy. If the energy of the SLAC electron–positron collision ring SPEAR was tuned to precisely 3.095 GeV, the number of particle collisions observed increased by a factor of more than a hundred. There was no known phenomenon that could explain this kind of resonance at such a high energy.

Believers in the standard model quickly realized that the resonance the experimenters were seeing was a bound state of a charmed quark and an anticharmed quark. If the charmed quark had a mass of about half of 3.095 GeV, then at that energy when an electron and positron collide and annihilate, they could produce a pair of charmed and anticharmed quarks more or less at rest. The strong force between these two quarks will keep them bound together. The narrowness of the energy range of the resonance implies that the lifetime of this bound state is much longer than that normally associated with strongly interacting particles. One reason for this is that the quark and antiquark are bound together at a short distance at which the strong force is becoming weak. The weakness of this interaction partially accounts for the fact that the quark and antiquark don't annihilate each other very quickly. The discovery of the J/psi managed to give new impressive evidence both for QCD and asymptotic freedom of the strong interactions, as well as providing the charmed quark predicted by the electroweak theory. In addition, the charmed–anticharmed quark bound state was a fairly simple system, quite analogous to positronium (the bound state of an electron and its antiparticle the positron). It had a complex behavior, including the existence of various excited states and all sorts of decay modes. These could be predicted from the standard model and measured at SLAC, with good agreement between the model and experiment.

By 1975, the name "standard model" was in use, at least at Harvard by Glashow, Weinberg, and others, and by 1979 it was common usage, appearing regularly in the titles of research articles. After 1974, many different types of confirmation of the theory poured in from different sorts of experiments around the world. At SLAC it took until 1976 for experimentalists to observe conclusively particles with naked charm, i.e., in which the charmed quark has combined with a quark not carrying charm. Over the next few years, a large amount of data about many different particles containing a charmed quark became available, all in close agreement with standard-model predictions.

The asymptotic freedom calculation in QCD determines not just the scaling behavior observed in the deep inelastic scattering experiments, but some small deviations from this scaling behavior. These deviations were observed during the late 1970s, precisely as calculated. At higher energies, QCD predicts that when an electron and positron collide, they will produce each different quark a calculable fraction of the time, and these quarks (together with quanta of the strong Yang–Mills field, called gluons) will produce jets of particles coming out of the interaction region. These jets were first seen in 1979, again exactly as predicted. The observation of these jets is in a sense finally an observation of quarks, since the production of a single quark is what is at the origin of each jet.

The most distinctive prediction of the electroweak theory was that of the existence of massive analogues of the photon: the quanta of the SU(2) Yang–Mills field. There are three of these, and they form one charged particle–antiparticle pair (the W^+ and W^-) and one uncharged single particle (the Z^0). All three of these particles were first seen in 1983 at the new proton–antiproton collider at CERN, again with exactly the masses and decay modes predicted by the standard model. By 1989, the Z^0 had been extensively studied with electron–positron colliders at SLAC (the SLC) and at CERN (LEP), where it appears as a resonance with energy about 90 GeV. The width of the energy range at which the resonance occurs measures the number of possible particles into which the Z^0 can decay and agrees exactly with what was expected. This implies that there are no other particles that experience electroweak interactions that we don't know about, at least up to a mass of one-half that of the Z^0 (45 GeV).

The one somewhat unexpected discovery during these years was that there is a third, even higher mass generation of particles. The first of these to be discovered was a new lepton, called the tau (τ) particle. It behaves just like an electron or muon, but with a larger mass. Early evidence for its existence was found at the SPEAR collider at SLAC in late 1975, and confirmed over the next two years.

Once the lepton of a third generation had been found, the standard model predicted that there should be a corresponding pair of quarks with two new flavors. These are known as the bottom and top quarks (an alternative terminology, truth and beauty, was popular among some, but has now been abandoned). The bottom quark was first seen in an experiment at Fermilab in 1977, and the first evidence of the existence of the top quark was also found at Fermilab, in this case in 1994. The three generations observed so far each have a neutrino, of zero or very small mass. The measurement of the resonance width of the Z^0 implies that if there are further generations beyond the first three, they must be a bit different, at least in having a very massive neutrino, with mass greater than 45 GeV. So far, there is no evidence of such a thing.

Glashow, Salam, and Weinberg received the Nobel Prize in 1979 for their work on the electroweak theory and 't Hooft and Veltman were similarly recognized in 1999 for their work on the renormalization of Yang–Mills theory. Finally, in 2004, Gross, Wilczek, and Politzer were awarded the prize for their calculation of the asymptotic freedom of Yang–Mills theory, leading to the realization that QCD is the correct theory of the strong interaction.

It is rather surprising that the Nobel award for asymptotic freedom took so long to occur, since the discovery was made in 1973, and by the 1980s a large amount of confirming experimental evidence was at hand. One can speculate that there were several probable reasons for this, a couple of them nonscientific. After his work on asymptotic freedom as a graduate student, Politzer soon became a tenured professor at Caltech, but then did relatively little work in mainstream physics. In addition, Gross had a reputation among some of his colleagues for sharp elbows, and there were some who held this against him. More importantly, 't Hooft had actually done the same calculation earlier, and even mentioned it at a conference in 1972. He never wrote up the result and was unaware of its significance for explaining the SLAC results and indicating the correct strong interaction theory. His Nobel award in 1999 for his related work cleared this obstruction to the award to Gross, Wilczek, and Politzer.

FURTHER READING

The Hunting of the Quark,[1] by Michael Riordan, contains a firsthand account from a SLAC experimentalist of some parts of the story covered in this chapter. *Nobel Dreams*,[2] by science writer Gary Taubes, is an entertaining (but rather unfair to its main subject, Carlo Rubbia) description of the discovery of the W and Z particles at CERN. Two excellent books about the standard model written by theorists involved in creating it are *In Search of the Ultimate Building Blocks*,[3] by Gerard 't Hooft, and *Facts and Mysteries in Elementary Particle Physics*,[4] by Martin Veltman.

The 2004 Nobel lectures[5] give more details about the discovery of asymptotic freedom and its later experimental confirmation.

8

Problems of the Standard Model

The standard model has been such an overwhelming success that elementary particle physics is now in the historically unparalleled situation of having no experimental phenomena to study that are in disagreement with the model. Every particle physics experiment that anyone has been able to conceive and carry out has given results in precise agreement with the standard model. There are, however, a small number of questions to which the standard model does not provide an answer, but about which one would expect a truly fundamental model to have something to say. These questions go more or less as follows:

- Why SU(3) × SU(2) × U(1)? A truly fundamental theory should explain where this precise set of symmetry groups is coming from. In addition, whereas QCD (the SU(3) part of this) has the beautiful property of having no free parameters, introducing the two other groups SU(2) and U(1) introduces two free parameters, and one would like some explanation of why they have the values they do. One of these is the fine structure constant α, and the question of where this number comes from goes back to the earliest days of QED. Another related concern is that the U(1) part of the gauge theory is not asymptotically free, and as a result it may not be completely mathematically consistent.

- Why do the quark and leptons of each generation come in a certain pattern? In mathematical terms, the quarks and leptons come in certain representations of the SU(3) × SU(2) × U(1) symmetry group. Why these specific representations and not others? This includes the question of why the weak interactions are chiral, that is, with particles of only one handedness experiencing the SU(2) gauge-field force.

- Why three generations? Could there be more with higher masses that we have not seen?
- Why does the vacuum state break the electroweak gauge symmetry? If the origin of this really is a Higgs field, then at least two new parameters are needed to describe the size of the symmetry breaking and the strength of the Higgs interaction with itself. Why do these parameters have the values they do? One of these parameters is determined by the observed properties of the electroweak interactions, but the other is still undetermined by any experimental result. This is why the standard model predicts the existence of a Higgs particle, but does not predict its mass. In addition, the standard quantum-field-theory description for a Higgs field is not asymptotically free, and again, one worries about its mathematical consistency.
- What determines the masses and mixing angles of the quarks and leptons in the theory? These particles have a pretty random-looking pattern of masses, giving nine numbers that the theory doesn't predict and that therefore have to be put in by hand. The mixing angles are four more parameters that determine precisely how the electroweak forces act on the particles. In the standard model, these thirteen parameters appear as the interaction strengths of the Higgs field with the quarks and leptons and are completely arbitrary. This problem is closely related to the previous one, since our inability to predict these parameters is probably due to not understanding the true nature of the electroweak gauge symmetry breaking of the vacuum.
- Why is the θ parameter zero? This parameter determines the size of a possible extra term in the QCD part of the theory, one that experiments show to be absent. One would like some explanation of why this term doesn't appear, or equivalently, why this parameter is zero or at least very small.

One way of thinking about what is unsatisfactory about the standard model is that it leaves seventeen nontrivial numbers still to be explained, and it would be nice to know why the eighteenth one is zero. Of the seventeen, fifteen show up in the standard model as parameterizing the properties of the Higgs field. So most of our problem with the standard model is to find a way either to get rid of the Higgs field or to understand where it comes from. Glashow, whose early version of the electroweak theory was incomplete (unlike the later Weinberg–Salam model) because it lacked something like the Higgs field to break the gauge symmetry, has been known to refer to the Higgs field as "Weinberg's toilet." His analogy is that the Higgs field is like the toilet one has in one's home: it fulfills an important and necessary function, but one is not proud of it and doesn't want to show it off to the neighbors.

One complication that has been ignored so far involves neutrinos. In the simplest version of the standard model all neutrinos are massless. Recent experiments have produced convincing evidence that this is not the case, although the experimental results are not quite good enough yet to determine completely the masses and mixing angles. Evidence for neutrino masses comes from the fact that neutrinos are observed to oscillate between different types. This phenomenon was first observed when experiments designed to detect neutrinos from the sun found only a third of the expected number. The explanation for this is that the electron neutrinos produced in the sun oscillate as they travel to Earth, sometimes appearing as muon or tau neutrinos, to which the experiments were not sensitive. The Fermilab NUMI/MINOS experiment that has just begun is one of several that should provide a more detailed understanding of this oscillation phenomenon.

It is relatively easy to extend the standard model in a simple way that allows for neutrino masses. This introduces a new set of seven parameters very much like the quark masses and mixing angles. The situation is slightly more complicated, since the fact that the neutrino has no charge allows two different sorts of mass terms. The exact mechanism responsible for these masses and mixing angles is just as mysterious in the neutrino case as it is for the quarks.

There is one remaining important part of physics that is completely ignored by the standard model: the gravitational force. This force is governed by Einstein's theory of general relativity, and the strength of its action on a particle is just proportional to the particle's mass. This constant of proportionality is Newton's constant, which is something that a truly fundamental theory should be able to calculate. The gravitational force is extraordinarily weak, and the only reason we are aware of it is that it is always attractive, never repulsive, so the forces from all the particles in the earth add up to produce a sizable force of gravity on everything at the earth's surface.

The weakness of the gravitational force is such that all of its observed effects can be understood and calculated without using quantum mechanics. Still, there is a gravitational field, and for consistency with the rest of physics, one would like to treat it using quantum field theory. The quantum of this field, the graviton, would interact so weakly with everything else as to be completely unobservable by any present or conceivable experiment. If one applies the standard method of perturbation expansion of a quantum field theory to general relativity, it turns out that the quantum field theory one gets is nonrenormalizable. This is the problem of quantum gravity: how does one find a consistent quantum theory for which general relativity is a good classical physics approximation? What makes this problem especially hard is that one has no experimental guidance about where to look for a solution, and no way of checking the kinds of predictions that tend to come out of any conjectural theory of quantum gravity.

Beyond the Standard Model

The discovery of asymptotic freedom in the spring of 1973 brought to completion the set of ideas needed for the standard model, and also brought to a close an exciting period of dramatic progress in our understanding of particle physics. What none of the theorists involved in this excitement could have guessed is that it was also the beginning of what was to be an exceedingly frustrating new era, one that has lasted more than thirty years to the present day.

Almost immediately after the advent of QCD, particle theorists began exploring new ideas that they hoped would address the problems left open by the standard model. As mentioned in Chapter 7, by 1975, the term "standard model" was already in use, and the terminology was meant to refer not just to its successes, but also to its role as a baseline for future progress. The main research programs that would dominate the next decade were all in place by 1975, and this chapter will consider each of them in turn.

GRAND UNIFIED THEORIES

In 1974, Glashow, together with a Harvard postdoc, Howard Georgi, created the first of a class of generalizations of the standard model that were to become known as grand unified theories and acquire the acronym GUTs. The idea of these models was to put together QCD and the electroweak theory into one single gauge theory using a larger group of symmetries, thus the term "grand unification." The original Georgi–Glashow model used the symmetry group SU(5), the group of special unitary transformations on five complex variables. Three of these five complex variables were those of QCD, and two were those of the electroweak theory. The use of all five variables at once meant that there were now symmetry transformations that related the quarks and the leptons.

The use of a single group like SU(5) just changed the question, "why SU(3) × SU(2) × U(1)?" to "why SU(5)?" but in the process it promised the possibility of calculating two of the standard-model parameters, those that give the relative strengths of the forces corresponding to the three groups in the standard model. Instead of the somewhat complicated pattern of representations of SU(3) × SU(2) × U(1) occurring in one generation of leptons and quarks, Georgi and Glashow were able to fit the particles of one generation precisely into just two representations of the symmetry group SU(5). The model had nothing to say about why there were three generations.

The SU(5) GUT also had nothing to say about the Higgs particle or the mechanism for vacuum symmetry breaking, and in fact, it made this problem much worse. The vacuum state now needed to break not just the electroweak symmetry, but much of the rest of the SU(5) symmetry. Another set of Higgs particles was required to do this, introducing a new set of undetermined parameters into the theory. Whereas the energy scale of the electroweak symmetry breaking is about 250 GeV, the SU(5) symmetry breaking had to be at the astronomically high energy scale of 10^{15} GeV. This means that most of the characteristic predictions of the theory have no hope of being seen in particle collisions at foreseeable accelerator energies.

There is one prediction that the SU(5) GUT makes that allows it to be tested by experiment. Since its gauge symmetry relates quarks and leptons, a quark inside a proton can turn into a lepton, causing the proton to disintegrate. The rate for this is very slow, because it is being caused by quanta of the SU(5) gauge field that, due to vacuum symmetry breaking, have extremely large masses. One can calculate the decay rate to find that the theory predicts that the average lifetime of a proton is about 10^{29} years. This is far greater than the age of the universe (about 10^{10} years), so one cannot expect to look at one proton and wait for it to decay. On the other hand, one can look at 10^{29} protons for one year to see whether any of them decay. Soon after 1974, several experimental groups began planning such experiments, which were conducted by putting detectors in and around a large volume deep underground in order to shield the experiment as much as possible from cosmic rays. I recall an experimentalist, Carlo

Rubbia, describing such an experiment as "just put half a dozen graduate students a couple of miles underground to watch a large pool of water for five years."

By the early 1980s, several such experiments were gathering data and soon were able to show that the SU(5) theory had to be wrong. The lifetime of the proton can't be 10^{29} years. Instead, it is at least 10^{31} to 10^{33} years, depending on what assumptions one makes about exactly what it will decay into. The SU(5) theory had been shown to be wrong, in a classic example of the scientific method. It made a specific prediction that was checked and found to be incorrect, falsifying the theory.

The SU(5) GUT was only the first of a large class of such theories that many people worked on after 1974. Other such GUTs all involve even larger symmetry groups and have the same problems with symmetry breaking as the SU(5) theory. These theories were intensively investigated during the late 1970s and early 1980s. A yearly conference on the subject of GUTs was held from 1980 to 1989, but the organizers decided in 1989 that the tenth of these conferences was to be the last. The fact that experiment ruled out the simplest of the GUTs and provided no positive evidence for the more complicated ones meant that other ideas were necessary.

TECHNICOLOR

Most of the problems of the standard model come from the introduction of the Higgs field and the associated arbitrariness in how it interacts with all the other elementary fields. Since the Higgs field itself has never been observed, it is very tempting to try to find some other mechanism that can play its role. What is needed is some mechanism that will cause the vacuum state not to be invariant under the SU(2) gauge symmetry transformations and thus have spontaneous symmetry breaking, while at the same time somehow giving masses to the fermions.

Another potential way of doing this is to find some other particles and forces such that the lowest-energy state breaks the SU(2) symmetry. This is called "dynamical" spontaneous symmetry breaking and is the sort of thing that happens in the case of superconductivity.

Recall from the discussion of spontaneous symmetry breaking that the theory of superconductivity was historically the first case in which spontaneous breaking of a gauge symmetry was considered. In a superconductor there is no elementary field to cause the symmetry breaking, but the dynamics of how electrons interact with a solid as they move through it leads to the lowest-energy state not being invariant under the gauge symmetry. So the superconductor is an example of dynamical spontaneous symmetry breaking, and one would like to see whether the same sort of thing could happen in the standard-model quantum field theory.

The standard vacuum state in quantum field theory as studied in a perturbation series is invariant under the symmetries of the theory, so in the perturbation series approximation there is no dynamical symmetry breaking. This makes dynamical symmetry breaking hard to study, since the simplest calculational methods won't work. One kind of nonperturbative force that will produce a noninvariant vacuum state is the strong force between quarks in QCD. Recall that in QCD there are two approximate SU(3) symmetries mixing the three low-mass quark flavors, one of which is spontaneously broken by the strong dynamics of the theory.

From the earliest days of the standard model there has been much interest in schemes for dynamical symmetry breaking. In 1978, Steven Weinberg and Leonard Susskind independently came up with the same proposal for a model of how this could happen. They suggested that there might be another unknown strong force very similar to QCD, governed by a gauge theory exactly like QCD, but with a different symmetry group. Since the QCD charges had been called colors, this theory acquired the name "Technicolor." The idea was that there could exist Technicolored particles, much like quarks, that are bound together by the forces of this new gauge theory into particles much like the mesons, nucleons, and other known particles of the strong interactions. What Weinberg and Susskind showed was that if the analogue of the spontaneously broken SU(3) in current algebra were broken in the Technicolor theory in the same way that it was in QCD, this would cause a dynamical spontaneous breaking of the gauge symmetry of the weak interactions, and one would have no need of a Higgs field. The lowest-energy mesons of the Technicolor theory would play the role that the Higgs field had played in the standard Glashow–Weinberg–Salam electroweak gauge theory.

Unfortunately, it soon became clear that there were various problems involved in making the Technicolor scheme work out in a way that agreed with experiment. The quarks and leptons are supposed to get their masses from their interaction with the Higgs field, and if one removes the Higgs field, one needs a new mechanism for producing these masses. To make this work out correctly with Technicolor, it was necessary to introduce yet another new set of forces and associated gauge fields, and these were given the name "extended Technicolor." By now, the proposed theory had become rather complicated and was postulating a large number of unobserved particles. All of these new particles and forces were strongly interacting, so there was no reliable way of calculating exactly what their effects would be. The theory was thus unable to make any accurate predictions and seemed to require too many complicated additions just to get it in qualitative agreement with the properties of the observed world. By the late 1980s, it was being studied by a dwindling number of physicists.

If anything like the Technicolor idea is true, once accelerator energies get high enough for one to be able to see what happens at the energy scale of dynamical electroweak symmetry breaking (250 GeV), experiments should be able to see quite different behavior from that predicted by the standard model with the Higgs field. The Large Hadron Collider at CERN should have enough energy for this, so by the year 2008, perhaps, experiments will start providing some guidance as to whether dynamical symmetry breaking is really what is going on and, if so, some hints as to what is causing it.

SUPERSYMMETRY AND SUPERGRAVITY

A couple of years before the standard model was in place, independent groups of Russian physicists in Moscow and in Kharkov published papers on an idea for a new sort of symmetry that was soon given the name "supersymmetry." The work of Evgeny Likhtman and Yuri Golfand in 1971, and Vladimir Akulov and Dmitri Volkov in 1972, was soon joined in 1973 by another independent version of the idea coming from two physicists at CERN, Julius Wess and Bruno

Zumino. The work of the Russian physicists was largely ignored at the time, and Golfand was soon forced out of his job in Moscow for some combination of reasons including being unproductive, being Jewish, and being politically unreliable. The work of Wess and Zumino, on the other hand, received considerable attention and fell on very fertile ground, coming just a few months after the discovery of asymptotic freedom, when many people were looking for new ideas about how to go beyond the new standard model.

A rough description of the basic ideas of supersymmetry follows. For more details one can consult an enthusiastic recently published popular book: *Supersymmetry: Unveiling the Ultimate Laws of Nature*, by the physicist Gordon Kane.[1]

Recall that any sort of quantum theory is supposed to have fundamental space-time symmetries corresponding to translations and rotations in four-dimensional space-time. To each symmetry there is a quantum-mechanical operator on Hilbert space that determines how an infinitesimal symmetry transformation affects the state. Translations in the three space dimensions correspond to the three components of the momentum operator, and translations in the time direction to the energy operator. Rotations in three-dimensional space correspond to the angular momentum operator, which again has three components. "Boosts," or transformations that mix space and time coordinates while keeping the light cone invariant, are the final operators to consider. The symmetry group that includes all of the space-time symmetries is called the Poincaré group, after the French mathematician Henri Poincaré. In mathematics, an algebra is basically just a collection of abstract objects with a rule for consistently multiplying and adding them. The operators that generate infinitesimal space-time symmetries form an algebra called the Poincaré algebra. To any Lie group of symmetry transformations one can associate an algebra of infinitesimal symmetry transformations. This is called a "Lie algebra," and the Poincaré algebra is just the Lie algebra associated with the Poincaré group.

What the Russian physicists had found was a consistent way of extending the Poincaré algebra of infinitesimal space-time symmetries by adding some new operators. The new, extended algebra was called a supersymmetry algebra. The new operators were in some sense square roots of translations: one could multiply two of them and get the momentum or energy operator that gives infinitesimal translations in space or time. Doing this required using the Clifford algebra that Dirac had rediscovered for his "Dirac equation."

If one tries to build these new operators in a quantum field theory, they have the peculiar feature that they relate bosons and fermions. Bosons, named after the Indian physicist Satyendra Nath Bose, are particles that behave in the simplest way that one might expect. Given two identical bosons, if one interchanges two of them, the quantum state stays the same. One can put as many of them as one wants in a single quantum state. It is a fundamental fact about quantum field theories, first shown by Pauli, that particles with an integer spin must be bosons. The only fundamental particles that have been observed that are bosons are the quanta of gauge fields: the photon for electromagnetism and its analogues for the strong and weak forces. These particles have spin quantum number 1. Hypothetical particles that would be bosons include the Higgs particle (spin 0) and the graviton (spin 2).

There is another consistent possibility for the behavior of identical particles in quantum theory, and particles that implement it are called fermions, named after Enrico Fermi. Fermions have the property that if one interchanges two identical particles, the state vector stays the same except for acquiring a minus sign. This implies that one cannot put two of these identical particles in the same state (because the only state vector that is equal to its negative is the zero vector). Pauli also showed that in a quantum field theory fermions will have half-integral spin, and examples include the known leptons and quarks, all of which are fermions with spin one-half.

The new operators that extend the Poincaré algebra transform fermions into bosons and vice versa, thus forcing a certain relation between the bosons and fermions of a quantum field theory with this new kind of symmetry. This gets around the Coleman–Mandula theorem of 1967, which implied that the only way one could get new symmetries besides the space-time ones was by using purely internal symmetries. The Coleman–Mandula theorem had implicitly assumed that the symmetry was taking bosons to bosons and fermions to fermions, not mixing the two the way supersymmetry does.

If a quantum field theory has supersymmetry, then for every bosonic or fermionic particle in the theory there must be another one of the opposite kind and with spin differing by one-half. The main reason that no one was very interested in the early Russian work was that there were no obvious pairs of particles in nature that could be related by this kind of symmetry. With the widespread acceptance of the standard model, the attention of theorists turned to looking for new sorts of symmetries and ways to relate parts of the standard model that were currently unrelated. Supersymmetry seemed a promising avenue to investigate, despite the absence of any evidence for its occurrence.

Another reason for being interested in supersymmetry was the hope that it might help with the problem of constructing a quantum field theory for gravity. One of the main principles of general relativity is what is called "general coordinate invariance," which means that the theory doesn't depend on how one changes the coordinates one uses to label points in space and time. In some sense, general coordinate invariance is a local gauge symmetry corresponding to the global symmetry of space and time translations. One hope for supersymmetry was that one could somehow make a local symmetry out of it. This would be a gauge theory and might give a new version of general relativity, preferably one whose quantum field theory would be less problematic.

The year after the first work on supersymmetry at Kharkov in 1972, Volkov together with Vyacheslav Soroka began working on the problem of finding a gauged version of supersymmetry that could provide a theory of gravity, and published some partial results. The development of what came to be called supergravity turned out to be a complex task, and a complete version of such a theory was finally written down only in the spring of 1976 by Daniel Freedman, Peter van Nieuwenhuizen, and Sergio Ferrara. Over the next few years, many physicists were involved in the project of developing good calculational tools for this theory and studying the problem of whether it was renormalizable. Because of supersymmetry, supergravity had to contain not just a graviton, which is a spin–2 boson, but also a spin-$\frac{3}{2}$ fermion called the gravitino. The hope was that cancellations in the perturbation series calculation between contributions from the graviton and those from the gravitino would allow one to avoid the infinities that led to the nonrenormalizability of the standard theory of quantum gravity. This was shown to happen to some extent, but not completely. If one went to high enough order in the perturbation series calculation, the infinities should still be there.

Besides the simplest supergravity theory, many physicists also began studying more complex theories, called extended supergravity theories, in which there are several different supersymmetries at once. These theories included not only a graviton and gravitino, but also any of a wide array of other particles and forces, perhaps even enough to include everything in the standard model. It turned out that one could hope to define consistently theories with up to eight different supersymmetries. The reason for this limit is that one can show that a theory with more than eight supersymmetries must contain elementary particles with spin greater than two, and such theories are believed to be physically inconsistent. The theory with eight supersymmetries was called "$N = 8$ extended supergravity," and was very popular for a while. For his inaugural lecture as Lucasian professor at Cambridge on April 29, 1980, Stephen Hawking chose the title "Is the end in sight for theoretical physics?" and argued that particle theory was very close to having found a complete and unified theory of all the forces in physics.[2] His candidate for such a theory was $N = 8$ extended supergravity, and he expressed the opinion that there was a fifty-fifty chance that by the end of the century, 20 years later, there would be a successful fully unified theory.

Many at the time would have predicted that Hawking would not be around (having succumbed to his difficult health problems) in the year 2000, but fortunately that is not the case. Alas, $N = 8$ extended supergravity has not fared as well. The first problem appeared soon after Hawking's lecture, when it was shown that at high enough order in the perturbation series, $N = 8$ extended supergravity was likely to continue to have the same nonrenormalizability problems as ordinary gravity. It also appeared that even with all the particles coming from the eight supersymmetries, there were not enough to account for all those of the standard model.

One way of constructing the $N = 8$ extended supergravity theory is to start by writing down the simplest supergravity theory with only one supersymmetry, but in eleven space-time dimensions. If one then assumes that for some unknown reason everything really depends on only four of the eleven dimensions, one can get the $N = 8$ extended theory in those four dimensions. This idea of trying to get a unified theory by thinking about physical theories with more than four dimensions of space and time goes all the way back to 1919. At that time, the mathematician Theodor Kaluza discovered that he could derive a theory containing both electromagnetism and gravity if he used Einstein's then new theory of general relativity but assumed that there were five dimensions of space and time, one of which was wrapped up everywhere as a small circle. The idea was that the extra dimension was so small that you couldn't see it, except indirectly through the existence of the electromagnetic force. Theories with this kind of extra dimensions became known as Kaluza–Klein theories, and were studied off and on for many years.

The Kaluza–Klein idea as applied to supergravity was to start with supergravity in eleven dimensions, and then to assume that, for some unknown reason, seven of the dimensions wrap up into something very small. For each point in four-dimensional space-time, the geometrical figure formed by the seven wrapped-up dimensions may have some symmetries, and one can try to interpret these symmetries as gauge symmetries. There may be enough of these extra symmetries to get the SU(3) × SU(2) × U(1) ones needed for the standard model. Soon, it was shown that there is a fundamental problem with this idea: one will always get a theory that is symmetric under mirror reflection, and thus can never get the weak forces in this way, since they are not symmetric under mirror reflection. The theory also continues to have the nonrenormalizability problem. For these reasons, by 1984, $N = 8$ extended supergravity was no longer so enthusiastically investigated, and physicists were beginning to become discouraged about supergravity in general as a way of solving the quantum gravity problem. Despite its problems, this 11-dimensional theory has recently been revived in a new context, one that will be described in detail in a later chapter.

10

New Insights in Quantum Field Theory and Mathematics

The physicist responsible for coming up with the elegant mathematical argument showing that Kaluza–Klein versions of supergravity could not explain the weak interactions was Edward Witten. This chapter will cover some of the progress that was made toward a better understanding of quantum field theory in the years following 1973, and Witten plays a role in all of this, the importance of which is hard to overemphasize. While learning quite a bit more about the physical aspects of quantum field theory, Witten and others have explored a host of new connections between these theories and mathematics, often bringing exciting new ideas and perspectives into already well developed areas of mathematical research. The discussion here will be at times unavoidably somewhat technical, but the hope is to give at least some of the flavor of how mathematics and physics research have been interacting at the most advanced levels.

EDWARD WITTEN

Edward Witten was born in Baltimore, Maryland, in 1951, the son of physicist Louis Witten, whose specialty is general relativity. As an undergraduate at Brandeis his interests were mostly nonscientific, and he majored in history and minored in linguistics. In 1968, at the age of 17, he published an article in the *Nation* about the New Left's lack of a political strategy, and another in the *New Republic* a year later about a visit to a commune in Taos, New Mexico. Witten graduated from Brandeis in 1971, spent a short time as a graduate student in economics at the University of Wisconsin, and worked for a while on the ill-fated McGovern presidential campaign of 1972. After deciding that politics was not for him, Witten entered the graduate program in applied mathematics at Princeton in the fall of 1973, soon transferring to the physics department. This was just after the discovery there earlier that year of asymptotic freedom by David Gross and his graduate student Frank Wilczek.

Witten's talent for theoretical physics was quickly recognized. A physicist who was a junior faculty member there at the time jokingly told me that "Witten ruined an entire generation of Princeton physics graduate students." By this he meant that it was a profoundly intimidating experience for them to see one of their peers come into graduate school without even a physics undergraduate degree, master the subject in short order, and soon start on impressive research work. Introducing Witten recently at a colloquium talk in Princeton,[1] my thesis advisor, Curtis Callan, Jr., recalled that Witten was a source of frustration to his thesis advisor, David Gross. Gross was convinced that the only way really to learn physics was to do calculations, and he kept giving new problems to Witten to work on, problems that he thought would require doing a complicated calculation. In all cases Witten would soon return with the answer to the problem, having found it from the use of general principles, without having had to do any calculation. Witten's first research paper was finished in late 1975. At the time of this writing, 311 more have appeared.

After receiving his PhD from Princeton in 1976, Witten went to Harvard as a postdoc and later a junior fellow. His reputation began to spread widely, and it was clear that a new star in the field had appeared. I gratefully recall his willingness to take time to help one undergraduate there who was trying to learn Yang–Mills quantum field theory (despite it being way over his head). In 1980, he returned to Princeton as a tenured professor, having completely bypassed the usual course for a particle theorist's career, which normally includes a decade spent in a second postdoc and a tenure-track assistant professorship. The fact that Harvard did not match Princeton's offer and do everything possible to keep him there is widely regarded as one of the greatest mistakes in the department's history. Witten moved across town to a professorship at the Institute for Advanced Study in 1987, and has been there ever since, with the exception of two years recently spent as a visiting professor at Caltech. He is married to another particle theorist, Chiara Nappi, who is now on the faculty at Princeton.

The MacArthur Foundation chose Witten in 1982 for one of its earliest "genius" grants, and he is probably the only person that virtually everyone in the theoretical physics community would agree deserves the genius label. He has received a wide array of honors, including the most prestigious award in mathematics, the Fields Medal, in 1990. The strange situation of the most talented person in theoretical physics having received the mathematics equivalent of a Nobel Prize, but no actual Nobel Prize in physics, indicates both how unusual a figure Witten is, and also how unusual the relationship between mathematics and physics has become in recent years.

When I was a graduate student at Princeton, one day I was leaving the library perhaps thirty feet or so behind Witten. The library was underneath a large plaza separating the mathematics and physics buildings, and he went up the stairs to the plaza ahead of me, disappearing from view. When I reached the plaza he was nowhere to be seen, and it is quite a bit more than thirty feet to the nearest building entrance. While presumably he was just moving a lot faster than I was, it crossed my mind at the time that a consistent explanation for everything was that Witten was an extraterrestrial being from a superior race who, since he thought no one was watching, had teleported back to his office.

More seriously, Witten's accomplishments are very much a product of the combination of a huge talent and a lot of hard work. His papers are uniformly models of clarity and of deep thinking about a problem, of a sort that very few people can match. Anyone who has taken the time to try to understand even a fraction of his work finds it a humbling experience to see just how much he has been able to achieve. He is also a refreshing change from some of the earlier generations of famous particle theorists, who could be very entertaining, but at the same time were often rather insecure and not known always to treat others well.

INSTANTONS IN YANG–MILLS THEORY AND IN MATHEMATICS

During the years since the standard model reached its final form, one of the main themes in research in particle theory has been the continuing effort to develop methods of calculation in quantum field theory that go beyond that of the perturbation expansion. This is of great importance in QCD, where the interaction becomes strong at large distances and so the perturbation expansion is useless. While progress of many kinds has been made, there still remains no fully successful nonperturbative calculational technique that can be applied to QCD.

In contrast to electromagnetism, in Yang–Mills theory not only is the quantum field theory highly nontrivial, but so are the classical field equations. In the absence of any charged particles, the differential equations that govern classical electromagnetism, Maxwell's equations, are very easily solved, and the solutions describe electromagnetic waves. These equations are linear, which means that if you have two solutions to the equations, you can add them and get a third. The analogous differential equations of classical Yang–Mills theory, called the Yang–Mills equations, are another story. They are nonlinear and quite difficult to solve explicitly. In 1975, four Russian physicists (Alexander Belavin, Alexander Polyakov, Albert Schwarz, and Yuri Tyupkin) were studying the Yang–Mills equations and found a way of getting at least some of their solutions, those that satisfy a condition called self-duality. These solutions became known as the BPST instantons. The name instanton refers to the fact that these solutions are localized around one point in four-dimensional space-time: an instant. An important technicality is that these solutions are for a so-called Euclidean version of the self-duality equations, where one treats time and space on a completely equal footing, ignoring the distinguishing feature that makes the time direction different in special relativity. These instanton solutions to the self-duality equations were something that mathematicians had never really thought much about.

One aspect of these instanton solutions was that they provided different starting points for a perturbation expansion calculation. The standard perturbation expansion can be thought of as an approximation that is good for fields close to zero, but one can develop a similar perturbation expansion starting not with zero fields, but with fields that solve the self-duality equations. Taking into account all perturbation expansions about all solutions should be a better approximation to the full theory than just using the standard perturbation expansion about the zero field. Calculations of this kind that use nontrivial solutions of the classical field equations as a starting point for an approximate calculation in the quantum theory are described as semiclassical. For the case of BPST instanton classical solutions, such calculations were performed by 't Hooft in 1976 and by several other groups of physicists soon thereafter. The results were of great interest in that they revealed new physical phenomena that did not occur in the standard perturbation expansion about zero fields. For example, 't Hooft found that this kind of calculation using an instanton solution of the classical equations for the electroweak theory led to a prediction of proton decay. The rate predicted was far slower than that predicted by grand unified theories, and also far slower than could ever possibly be measured, so was of purely theoretical interest. 'T Hooft also found that one of the unexplained features of current algebra, the nonexistence of a ninth Nambu–Goldstone boson, could potentially also be explained by these new calculations.

Over the next few years, a wide array of semiclassical calculations was performed using various solutions to the classical equations of both Yang–Mills theory and other physical theories. For a while there were high hopes, especially at Princeton, that this work might lead to a calculational method that would allow for reliable calculations in strongly interacting QCD. In the end, these hopes were not to be realized. It seems that semiclassical calculations still rely too much on forces being weak and, like the standard perturbation expansion, fail at the point that the forces in QCD become strong.

While instantons did not help much with understanding QCD, in a surprising turn of events they ended up being of great importance in mathematics. By any measure, one of the leading mathematicians of the second half of the twentieth century is Sir Michael Atiyah, whose work has been mainly in topology and geometry, although cutting across the standard boundaries of mathematical fields in a very unusual way. Probably his greatest achievement is the so-called Atiyah–Singer index theorem, which he and Isadore Singer proved in the mid-1960s. This earned them both the Abel Prize in 2004, only the second one ever awarded. This prize was set up by the Norwegian government in 2001 to provide an equivalent of a Nobel Prize for mathematics, and the first one was awarded in 2003 to the French mathematician Jean-Pierre Serre.

The Atiyah–Singer index theorem tells one the number of solutions to a large class of differential equations purely in terms of topology. Topology is that part of mathematics that deals with those aspects of geometrical objects that don't change as one deforms the object (the standard explanatory joke is that a topologist is someone who can't tell the difference between a coffee cup and a doughnut). One important aspect of the index theorem is that it can be proved by relating the differential equation under study to a generalized version of the Dirac equation. Atiyah and Singer rediscovered the Dirac equation for themselves during their work on the theorem. Their theorem says that one can calculate the number of solutions of an equation by finding the number of solutions of the related generalized Dirac equation. It was for these generalized Dirac equations that they found a beautiful topological formula for the number of their solutions.

In the fall of 1976, Singer, who had heard about Yang–Mills theory when visiting Stony Brook, was lecturing at the Mathematical Institute in Oxford on the subject. Atiyah became interested, and they quickly realized that their index theorem could be applied in this case and that it allowed a determination of exactly how many solutions the self-duality equations would have. From 1977 on, Atiyah's research work was dominated by topics suggested from theoretical physics, so much so that the fifth volume of his collected works is entitled *Gauge Theories*.[2]

In the spring of 1977, Atiyah was visiting Cambridge, Massachusetts, which was a not uncommon occurrence.[3] However, unlike previous visits, on this occasion he was very interested in talking to physicists and was impressed greatly by a Harvard postdoc whom he met in Roman Jackiw's office at MIT: Edward Witten. He invited Witten to visit him at Oxford for a few weeks, a visit that occurred in December 1977 and was the beginning of more than a quarter century of interactions between the two, which have led to great progress in both mathematics and physics. During 1978, Witten worked on several ideas involving supersymmetry, including an idea for using it to solve the full Yang–Mills equations, not just the self-duality equations. During Witten's visit to Oxford, Atiyah put him in touch with a British physicist named David Olive, and this led to joint work on some remarkable duality properties of certain super-symmetric gauge theories. This topic has turned out to be of great significance, and important work on it continues to this day. This period was the beginning for Witten of what was to be a very long and deep involvement with modern mathematics, supersymmetry, and the relationship between the two.

Many different sorts of mathematical ideas grew out of this early work on the self-duality equations, and the most surprising results soon came not from Atiyah, but from a student of his at Oxford named Simon Donaldson. By 1982, Donaldson managed to prove a variety of powerful and unexpected theorems about the topology of four-dimensional space, using as a basic technique the study of the solutions of the self-duality equations for Yang–Mills theory. The 1950s and early 1960s had been a golden era for the subject of topology, and by the end of the 1960s, quite a bit was known. Topology of two-dimensional spaces (think of the surface of a sphere or dough-nut) is a very simple story. All two-dimensional spaces are character-ized by a single nonnegative integer, the number of holes. The surface of the sphere has no holes, the surface of the doughnut has one, and so on. A surprising discovery was that things simplified if one thought about spaces with five dimensions or more. In essence, with enough dimensions, there is room to deform things around in many ways, and, given two different spaces, the question of whether one can be deformed into the other is intricate but can be worked out. Three and four dimensions turned out to be the really hard ones, and progress in understanding them slowed dramatically.

The classification of spaces by their topology depends on what sort of deformations are allowed. Does one allow all deformations (as long as one doesn't tear things), including those that develop kinks, or does one insist that the space stay smooth (no kinks) as it is deformed? It was known by the late 1960s that for spaces of dimension two and five or more, the stories for the two different kinds of deformations were closely related although slightly different. What Donaldson showed was that in four dimensions these two different kinds of deformations lead to two completely different classification schemes. The fact that he did this using gauge theory and the solutions to differential equations (parts of mathematics that topologists viewed as far from their own) just added to the surprise. Donaldson was awarded the Fields Medal in 1986 for this work.

LATTICE GAUGE THEORY

Another very different calculational method called lattice gauge theory was proposed independently by Kenneth Wilson and Alexander Polyakov in 1974. The idea of lattice gauge theory is to make Yang–Mills quantum field theory well defined, independently of any perturbation expansion, by constructing the theory not for every point of space and time, but just on a finite regular "lattice" of points. Defining the theory on only a finite number of points, each separated by some finite distance from its neighbors, eliminates the problems with infinities that make quantum field theory so difficult. These problems then reappear as one tries to put more and more points in the lattice, making the spacing between lattice points smaller and smaller. What Wilson and Polyakov found was that one could easily do this kind of discretization of the theory in a gauge-invariant way if one associated fields describing quarks and leptons to the points of the lattice, and Yang–Mills fields to the links connecting neighboring points on the lattice.

There is a general technique for defining quantum field theories (due to Feynman) called the "path integral" technique. The name comes from the fact that if one uses this technique to define quantum mechanics (as opposed to quantum field theory), it involves doing the sort of integrals done in calculus, but now over the infinite-dimensional space of all paths or trajectories of particles in space and time. The path integral technique had been used mainly for rather formal calculations in quantum field theory, since in general no one knew how to do the infinite-dimensional integrals it led to. In lattice gauge theory the integrals become well-defined due to the discrete nature of the lattice. If one restricts attention to a finite part of space-time, and to a lattice of points in it, each separated by a finite distance, then the number of points is finite and the dimension of the integrals to be done is now finite. Of course, the calculation one really wants to do involves taking the lattice spacing to zero and going to the limit of an infinite number of points, but Wilson and

Polyakov knew from studying similar problems in condensed matter physics that there was a chance this could be done successfully.

While many techniques used by condensed matter physicists give some sort of insight into lattice gauge theory, most of them rely upon one kind or another of approximation scheme whose reliability in this particular case is in doubt. The exception is something called the Monte Carlo algorithm, which is a probabilistic calculational method for doing very-large-dimensional integrals of certain kinds. To approximate an integral over a large-dimensional space, the Monte Carlo algorithm works by having a computer generate points in the space randomly, with a probability of any point showing up being proportional to the value of the function one wants to integrate at that point. For certain classes of functions in high dimensions this works very nicely, giving one a fairly good approximation to the integral within a manageable amount of computer time. The longer one runs the calculation and the more points one generates, the better the result. As the calculation goes on, one can see whether it is converging to a well-defined result. The first calculations of this kind were done in 1979 by Michael Creutz at Brookhaven, and many groups continue to work in this area. Some of my colleagues in the physics department at Columbia are part of a group that has designed and built specialized multiprocessor computers to do this kind of calculation, and they now have machines capable of 10 teraflops, that is, 10^{13} floating-point calculations per second.

This kind of calculation works quite well in pure Yang–Mills theory without fermion fields, but calculations with fermions are more difficult, especially if one allows for the effects of particle–antiparticle pairs. The results of doing calculations in QCD by this method are consistent with experimental results about strongly interacting particles, but are still far from giving precision calculations of the sort that are possible in QED. While these calculations do provide strong evidence that QCD correctly describes the strong interactions, many theorists find them unsatisfying. Besides the problems in dealing with fermions, one ideally would like to be able to calculate things in QCD in some way that would count as an explanation of what is going on that a human being can comprehend. In the current lattice calculations this is far from being the case.

Large N

Another possible method for doing calculations in QCD was proposed by 't Hooft in 1974, soon after the discovery of asymptotic freedom. It is based on the idea of generalizing QCD from an SU(3) gauge theory involving three colors to one in which the number of colors is some arbitrary number N and the corresponding symmetry group is SU(N). The idea that 't Hooft put forward was that the theory may actually in some sense simplify as the number N increases, and one could hope to solve the theory exactly in the limit as N goes to infinity. Then one could further hope to construct a new sort of perturbation expansion, one in which the expansion parameter was $\frac{1}{N}$, the inverse of N. This has become known as a $\frac{1}{N}$, or large-N, approximation technique. The hope is that for the real-world QCD case of $N = 3$, the number $\frac{1}{3}$ may be a small enough number that knowledge of just the first term or two of the $\frac{1}{N}$ expansion would give results fairly close to the exact solution of QCD.

This idea can be made to work in simplified problems, especially ones in which the quantum field theory is not in four space-time dimensions but in two space-time dimensions (one dimension of space, one of time) instead. In these simplified "toy" models, the three dimensions of real space are replaced by a single spatial dimension. This new space is a single line, and the theory simplifies quite a bit. Witten took up the large-N concept in 1978, and over the next few years managed to use it to give some impressive qualitative arguments that this was a promising way of thinking about QCD. His initial work involved two-dimensional space-time toy models in which good calculational methods were available and one could study explicitly the expansion in powers of $\frac{1}{N}$. He later gave a convincing explanation of how the current-algebra model of pions should fit into the large-N expansion, in the process resolving the longstanding problem of explaining why there is no ninth low-mass Nambu–Goldstone boson.

The culmination of this line of work came in 1983, when Witten showed that not only could the physics of pions be understood from within current algebra, but so could the other much more massive strongly interacting particles such as the proton and the neutron. To

do this required thinking of the proton and neutron as exotic configurations of the pion fields that carried nontrivial topology. For topological reasons these field configurations could not be untwisted and
deformed into the small variations in the pion field whose quanta
were the pions. Witten used a tour de force combination of arguments about the probable behavior of the large-N approximation, together with some beautiful geometry and topology, to derive this
result. I remember going to his first public seminar about this, given
at the Institute for Advanced Study. The seminar began with a review
of the current-algebra picture, the large-N approximation, and a certain amount of topology. It climaxed when he stopped for effect at
the moment that his calculation had shown that the topology led to
the existence of a proton. It was rather like a magician pulling a rab-
bit out of the hat, and the audience was suitably impressed, not an
easy thing to accomplish with that kind of audience in that location.

Despite partial progress, neither Witten nor anyone else has yet
been able to find a way to use the large-N expansion to solve QCD.
The fundamental problem remains that no one knows how to write
down the exact solution of SU(N) gauge theory in the limit as N goes
to infinity. This is the starting point, or zeroth-order term, for an expansion in powers of $\frac{1}{N}$, and without it one can't get off the ground
and start doing precise calculations. A longstanding conjecture is that
this large-N limiting theory is some sort of string theory, but exactly
how to make this idea work remains an open problem. Recent years
have seen extremely interesting progress in this area, which will be
discussed in a later chapter.

TWO-DIMENSIONAL QUANTUM FIELD THEORIES

Much of the new understanding of quantum field theory that has
been gained since 1973 has come from the study of quantum field
theories such as the large-N example mentioned earlier, in which the
three standard dimensions of space are reduced down to a single dimension. In these theories, space is a line, or perhaps a circle. A being in a world described by such a theory would have extension only
in one direction, and could only move backward or forward in that
direction. Theories of this kind share many of the characteristics of

quantum field theories with three spatial dimensions, but tend to be mathematically much more tractable. One reason for this is that the renormalization problems of quantum field theory are much simpler to deal with in one spatial dimension than in three, since several sources of infinities are absent. Besides the spatial dimension there is still a time dimension, so these quantum field theories are generally referred to either as $(1 + 1)$-dimensional or two-dimensional.

Theories of this kind were used in initial studies of semiclassical calculations using instantons, as well as in the large-N calculations. In each of these cases, one could see under which circumstances the approximate calculational technique in question would work out. This experience with these toy models was invaluable in gaining an idea of what would happen in the real four-dimensional calculations. During the 1980s, a great deal more was learned about these models, especially a subclass of them called conformal field theories, for which one could find exact solutions that do not depend on any approximation scheme.

The mathematics of two-dimensional surfaces is a beautiful and highly developed subject that goes back to Carl Friedrich Gauss in the early nineteenth century. This was investigated further by Bernhard Riemann and others later in the century, and their crucial insight is that it is a good idea to use complex numbers and to think of the surface as being parameterized not by two real numbers, but by one complex number. A surface parameterized in this way is called a Riemann surface. The study of Riemann surfaces is one of the most central topics in modern mathematics, since it stands at the intersection of several different fields, including topology, geometry, analysis, and even number theory.

A crucial concept in this subject is that of an analytic function, a concept for which it is hard to get an intuitive feel. Real-valued functions of one real variable are the subject of high school mathematics classes, and real-valued functions of two real variables are not too much harder to think about. In second-year calculus classes one uses them frequently, and often works with them by drawing their graphs, which are two-dimensional surfaces in three dimensions. If one generalizes this to let the function of two real variables take on complex rather than real values, one now has two real-valued functions, corresponding to the real and imaginary parts of the complex value. The ability to visualize the graph of the function is now lost, since it would take four dimensions to draw it.

If the two real variables on which the function depends are replaced by a complex variable, so that one is now considering complex-valued functions of a complex variable, then something dramatic happens. One can impose a new condition on the function that links together the complex nature of the domain and range of the function. Recall that the crucial thing that makes a plane parameterized by two real numbers into a complex plane is the specification of what it means to multiply by the square root of minus one. Geometrically, this is just a 90-degree rotation. The condition for a function to be analytic is that doing the 90-degree rotation on the domain of the function has the same effect as doing the rotation on the function's range.

This condition is still tricky to grasp, but one can show that another way to characterize such an analytic function is that such a function preserves angles. If one considers two line segments that join to form an angle somewhere in the domain, an analytic function will take these line segments to curves in the range that meet at exactly the same angle formed by the original line segments. So an analytic function gives one what is called a conformal transformation. This is a transformation of some part of the complex plane into another part that may change the sizes of things, but keeps angles the same. These conformal transformations form a group of symmetry transformations, and this group is infinite-dimensional, since it takes an infinite number of parameters to parameterize such symmetry transformations.

If one has a two-dimensional quantum field theory, one can consider how it behaves under these conformal transformations. A quantum field theory that is invariant or behaves simply under these transformations is called a conformal field theory. So conformal field theories are special kinds of two-dimensional quantum field theories, ones that have an infinite-dimensional group of symmetry transformations acting on them, a group of angle-preserving transformations. The first nontrivial theory with this property that was widely studied is called the Thirring model, after Walter Thirring, who first investigated it in 1958. In 1970, Polyakov, motivated by problems in condensed matter physics, began to investigate the general properties of conformal field theories, and a lot of work was done on these theories during the 1970s and 1980s. One important development was due to Witten, who in 1983 discovered what is now known as the Wess–Zumino–Witten model. His construction of this model used a two-dimensional version of the topological trick he had used to get protons out of current algebra, and the end result has a huge amount of fascinating mathematical structure built into it. Later work showed that a large class of conformal field theories are all related to the Wess–Zumino–Witten model, constructed out of it by adding in various different gauge symmetries.

While the theory of representations of finite-dimensional groups such as the ones Weyl studied in 1925–1926 was a well-developed part of mathematics by the 1960s, little was known about the representations of infinite-dimensional groups such as the group of conformal transformations in two dimensions. Without some restrictive condition on the groups and the representations to be considered, the general problem of understanding representations of infinite-dimensional groups appears to be completely intractable. The mathematicians Victor Kac and Robert Moody introduced some new algebraic structures in 1967 that allowed the construction of a class of infinite-dimensional groups, now known as Kac–Moody groups. These groups have some of the same structure as finite-dimensional ones, enough to allow a generalization of some of the techniques used by Weyl and others. One crucial formula found by Weyl is now known as the Weyl character formula, and it computes the "character" of a representation. This is a function defined on the group, a rule that gives a number for each group element. One can tell which representation of the group one has by computing this function, so it characterizes the representation. To tell whether two representations constructed by very different means are really the same, one simply needs to compute their character functions and see whether they are identical.

In 1974, Kac derived a character formula that was the generalization of Weyl's character formula to the case of his Kac–Moody groups, and this formula is now known as the Weyl–Kac character formula. Over the next few years, interest in these groups grew, and constructions were found for their representations. The methods used were a combination of generalizations of the finite-dimensional case together with techniques borrowed from physicists, including one technique involving something called a vertex operator. The vertex operator technique had been developed during the late 1960s and early 1970s as part of an attempt to understand the strong interactions using string theory, an effort that will be described in more detail later on. By 1974, the discovery of asymptotic freedom had caused most physicists to lose interest in this field, but a small number of mathematicians and physicists continued work in this area throughout the late 1970s and early 1980s.

In recent years, this work has led to the development of a field of mathematics now known as vertex operator algebra, involving new algebraic constructions inspired by conformal quantum field theories. This new field has had applications in a range of areas of mathematics, including some far removed from physics. Perhaps the best known of these applications has been to the study of representations of something known as the monster group. The monster is a group with a finite but extremely large number of elements (around 10^{55} of them). All groups with a finite number of elements can be decomposed into irreducible pieces, and the classification of these possible irreducible pieces was finally completed in a massive effort during the 1980s. The monster group is the largest possible irreducible piece of a finite group, and the fact that its representations can be understood using techniques ultimately derived from quantum field theory has been one of the more unexpected connections between mathematics and physics to turn up in the past few decades.

The Wess–Zumino–Witten two-dimensional quantum field theory turns out to be closely related to the representation theory of Kac–Moody groups. Just as the Hilbert space of quantum-mechanical models gives a representation of any finite-dimensional group of symmetry transformations of the model, the Wess–Zumino–Witten model has a symmetry group that is an infinite dimensional Kac–Moody group, and its Hilbert space is a representation of this group. The Hilbert space of the Wess–Zumino–Witten model is a representation not only of the Kac–Moody group, but of the group of conformal transformations as well (actually this is a serious over-simplification, but the Hilbert space can be decomposed into pieces for which this is true).

The Wess–Zumino–Witten theory has had a huge importance for both mathematicians and physicists. For mathematicians, it simultaneously provides an explicit construction of representations of two infinite-dimensional groups: the Kac–Moody group and the group of conformal transformations. The quantum-field-theoretical nature of the construction is something new to mathematicians, and it exposes the existence of a wealth of additional structure in these representations that had not been suspected before. For physicists, this is an interesting quantum field theory, one of very few that can be solved exactly. No perturbation theory arguments are needed, since one has access to the full, exact solution in terms of representation theory. The investigation of the implications for both mathematics and physics of this and other closely related quantum field theories continues to this day.

An intriguing aspect of this subject is that the Kac–Moody groups are examples of gauge groups, symmetry groups of the kind that proved to be so important in the development of the standard model. Unfortunately, they are groups of gauge symmetries corresponding to quantum field theories in two-dimensional space-time, not the four-dimensional space-time of the standard model. Whether the study of representations of these lower-dimensional gauge groups will give insight into the physics associated with gauge symmetries in four-dimensional space-time remains to be seen.

ANOMALIES AND QUANTUM-MECHANICAL SYMMETRY BREAKING

The name Wess–Zumino–Witten was attached to Witten's two-dimensional quantum field theory because it is a two-dimensional version of a four-dimensional model considered earlier by Julius Wess and Bruno Zumino. Wess and Zumino were studying the model because of its relation to current algebra and the theory, described earlier, of pions as Nambu–Goldstone particles. When Witten made the shift in attention from four dimensions to two, the analogue of the four-dimensional current algebra became essentially the mathematical structure studied by Kac and Moody. The study of the representations of Kac–Moody groups is closely related to this two-dimensional version of current algebra.

Early work on current algebra during the 1960s had turned up a rather confusing problem, which was dubbed an "anomaly." The source of the difficulty was something that had been studied by Schwinger back in 1951, and so became known as the problem of the Schwinger term appearing in certain calculations. The Schwinger term was causing the Hilbert space of the current algebra to be not quite a representation of the symmetry group of the model. The standard ways of constructing quantum-mechanical systems ensured that if there was a symmetry group of the system, the Hilbert space should be a representation of it. In the current algebra theory, this almost worked as expected, but the Schwinger term, or equivalently the anomaly, indicated that there was a problem.

The underlying source of the problem had to do with the necessity of using renormalization techniques to define properly the current-algebra quantum field theory. As in QED and most quantum field theories, these renormalization techniques were necessary to remove some infinities that occur if one calculates things in the most straightforward fashion. Renormalization introduced some extra U(1) phase transformations into the problem, ruining the standard argument that shows that the Hilbert space of the quantum theory should be a representation of the symmetry group. Some way needed to be found to deal with these extra U(1) phase transformations.

In two-dimensional theories, it is now well understood how to treat this problem. In this case, the anomalous U(1) phase transformations can be dealt with by just adding an extra factor of U(1) to the original infinite-dimensional symmetry group of the theory. The Hilbert space of the two-dimensional theory is a representation, but it is a representation of a slightly bigger symmetry group than one might naively have thought. This extra U(1) piece of the symmetry group also appears in some of the infinite-dimensional Kac–Moody groups. So in two dimensions the physics leading to the anomaly and the mathematics of Kac–Moody groups fit together in a consistent way.

In four-dimensional quantum field theories, the problem of the anomaly, or Schwinger term, is much trickier. Current algebra in four dimensions has led to a significant amount of understanding of the physical aspects of the problem. One of the earliest physical consequences of the anomaly concerned the rate at which neutral pions decay into two photons. If one ignores the anomaly problem, current algebra predicts that this decay will be relatively slow, whereas experimentally it happens very quickly. Once one takes into account the anomaly, the current-algebra calculation agrees well with experiment. This calculation depends on the number of colors in QCD, and its success was one of the earliest pieces of evidence that quarks had to come in three colors. Another successful physical prediction related to the anomaly was mentioned earlier. This is the fact that, ignoring the anomaly, there should be nine low-mass pions, the Nambu–Goldstone bosons of the spontaneously broken symmetry in current algebra. In reality, there are nine pions, but only eight of them are of relatively low mass. The higher mass of the ninth one can be explained once one takes into account the effect of the anomaly.

The anomaly phenomenon is sometimes called quantum-mechanical symmetry breaking, since the theory naively appears to have a certain symmetry, but the Hilbert space is not quite a representation of this symmetry, due to the subtleties of how quantum field theories are defined. In addition to the anomaly we have been discussing, which affects the global symmetries studied in current algebra, there can also be an anomaly in the gauge symmetry of a theory. This is called a gauge anomaly. Gauge anomalies are less well understood, but they definitely interfere with the standard methods for dealing with the gauge symmetry of Yang–Mills quantum field theory. If one throws out the quarks and considers the standard model with just the leptons, one finds that this theory has a gauge anomaly, and it ruins the standard renormalization of the quantum field theory as first performed by 't Hooft and Veltman. To this day, it is unknown whether there is some way around this problem, but it can be avoided, since if one puts the quarks back into the theory, one gets an equal and opposite gauge anomaly that cancels the one coming from the leptons. The full standard model has no gauge anomaly due to this cancellation, and the principle that gauge anomalies should cancel is often insisted on in considering any extension of the standard model.

During the early 1980s, there was a great deal of interaction between mathematicians and physicists interested in the anomaly problem. From the point of view of a mathematician, one aspect of the anomaly is that it is related both to the Atiyah–Singer index theorem and to a generalization known as the index theorem for families. Whereas the original index theorem describes the number of solutions to a single equation and does this in terms of the number of solutions of a Dirac equation, the families index theorem deals with a whole class, or family, of equations at once. A family of Dirac equations arises in physics because one has a different Dirac equation for every different Yang–Mills field, so the possible Yang–Mills fields parameterize a family of Dirac equations. This situation turns out to be one ideally suited to the use of general versions of the index theorem already known to mathematicians, and in turn has suggested new versions and relations to other parts of mathematics that mathematicians had not thought of before. As usual, Witten was the central figure in these interactions between mathematicians and physicists, producing a fascinating series of papers about different physical and mathematical aspects of the anomaly problem.

The two-dimensional version of this whole story has been exceptionally rich, and in recent years, interesting connections have been found between the representation theory of Kac–Moody groups and index theory. One version of this connection was first discovered by the physicist Erik Verlinde, in 1988. This discovery led to the formulation of something called a Verlinde algebra and to an associated Verlinde formula. The Verlinde formula tells one the number of solutions to a certain equation and is closely related to index theory, which it extends into a new mathematical area. The implications for mathematics of this entire set of ideas is still being actively explored.

While a great deal is known about the two-dimensional case, much about four dimensions remains a mystery. A fundamental reason for this is that in two dimensions things are determined by certain Kac–Moody groups whose representations are well understood. For groups of gauge symmetries in four dimensions, very little is known about the theory of their representations. Renormalization is much trickier in four dimensions, so the mathematical techniques that work in two dimensions fail here, and the necessary ones remain to be developed.

TOPOLOGICAL QUANTUM FIELD THEORY

While the anomaly story has had an important impact on mathematics, of greater importance has been a set of ideas that goes under the name topological quantum field theory. It all began with the publication in 1982 by Witten of an article with the title "Supersymmetry and Morse Theory." This article was published not in a physics journal, but in a prominent mathematics journal, the *Journal of Differential Geometry*. I have been told that at the time, the publication of this article was a subject of controversy among the editors. It is written in the style of a physics paper, ignoring the careful and precise sequence of definitions, theorems, and proofs that are characteristic of the mathematics literature. Its publication required some prominent mathematicians to exert their influence in its favor, and their judgment has been amply justified by what has come out of it.

The Harvard mathematician Raoul Bott recalled[4] giving some lectures about topology and the Yang–Mills equations to physicists at a summer school in Cargèse during the summer of 1979. Most physicists found Bott's lectures to be far from their concerns and took a detached view of them, but Witten was in attendance and paid close attention. Eight months later Witten wrote to Bott, "Now I finally understand Morse theory!"

Morse theory is a method for studying the topology of a space that goes back to the earliest history of the subject and work of the mathematician Marston Morse in 1925. What Witten had finally understood was a relation between Morse theory and quantum mechanics. He later showed in his article that for a given space of any dimension, he could construct a simple quantum-mechanical model with supersymmetry that had a Hilbert space that depended purely on the topology. This Hilbert space was finite-dimensional and corresponded exactly to something long known to mathematicians, the socalled homology of a space. His construction exploited a version of

Morse theory that, while it had been previously considered by some mathematicians, was not all that well known, even to the experts.

The long-term significance of Witten's results was not immediately clear. The homology of a space is one of the simplest examples of a general class of mathematical constructions called topological invariants. A topological invariant is something that doesn't change as one deforms a space, and thus depends only on its topology. A good topological invariant allows a topologist to tell whether two different spaces can be deformed into each other or are topologically distinct. Just compute the topological invariants of the two spaces, and if they are different, the spaces certainly cannot be deformed one into the other. A topological invariant associated with a space may simply be a number, but it can also be something more complicated. The homology invariant that Witten was dealing with is a set of integers, which in some sense tell one how many holes of different dimensions a space has.

The topological invariants Witten had found were all well known, but the quantum-mechanical methods he was using were unlike anything mathematicians had considered before. It appeared that Witten had a new and very powerful idea, and by extending it to more complicated quantum theories such as quantum field theories, there might be a lot of exciting new mathematics to be found. In his article Witten made a beginning on this, ending with some initial comments about two-dimensional quantum field theories.

In May 1987 a mathematics conference was held at Duke University to honor, two years late, the centenary of the birth of Hermann Weyl. This was the first mathematics conference I ever attended, and at the time I was finishing up my physics postdoc at Stony Brook, with no employment prospects for the coming academic year. By then I was pretty familiar with the world of theoretical particle physicists, but just starting to get to know a little about the mathematical community. The conference was taking place at a very exciting time for mathematics, with a lot of new ideas in the air. My experience there was an indication to me that if I could, it would probably be a good idea to change fields to mathematics.

Witten and Atiyah were at the conference, and Atiyah gave a talk on "New invariants of three- and four-dimensional manifolds"[5] that was the most remarkable talk I have ever heard. He described new topological invariants that his former student Donaldson had just defined for four-dimensional spaces, together with other new work by Andreas Floer, a brilliant young mathematician, who, sadly, was to take his own life a few years later. Floer's work was on topological invariants of three-dimensional spaces and he had defined a new topological invariant now called Floer homology using Witten's ideas about Morse theory. Atiyah traced out a conjectural picture of how Floer and Donaldson's ideas fit together. The basic idea involved thinking about a four-dimensional space with a three-dimensional boundary. As an analogy in one dimension lower, think of a three-dimensional ball, the boundary of which is the two-dimensional surface of the ball. Atiyah showed that the Floer homology of the three-dimensional boundary space was exactly what one needed to fix in order to make sense of Donaldson's new invariants in the case of a four-dimensional space with boundary. This neatly linked together the two new areas of mathematics that Donaldson and Floer had created, in the process suggesting a wealth of new questions.

Atiyah also related the whole picture to Witten's work on supersymmetry and Morse theory, suggesting that there should be a four-dimensional quantum field theory whose Hilbert space was the Floer homology of the boundary three-dimensional space, and whose observable quantities would be Donaldson's new topological invariants. Along the way he brought in a wide variety of other areas of mathematics and, as an aside, mentioned a topological invariant of knots called the Jones polynomial that had been discovered by Vaughn Jones in 1985. From what I recall, Witten was in the audience and maybe already knew about the Floer and Donaldson story, but after the talk went up to Atiyah and started asking him questions about the Jones polynomial.

At first, Witten was dubious that the quantum field theory whose existence was conjectured by Atiyah really existed.[6] After further prodding from another visit by Atiyah to the Institute for Advanced Study in late 1987, Witten returned to the question and soon found a quantum field theory with the properties Atiyah was looking for, publishing an article in early 1988 describing it entitled "Topological Quantum Field Theory." He had found this theory by starting with a four-dimensional quantum field theory with supersymmetry. Recall that supersymmetry is a symmetry that is in some sense a square root of translational symmetry. General four-dimensional spaces, especially ones that are topologically nontrivial, have a complicated and curved geometry, so they certainly do not have an overall translational symmetry, and thus not a standard supersymmetry. Witten was able to get around this with an ingenious trick, introducing what he called a twisted supersymmetry such that some of the supersymmetry still existed on a curved four-dimensional space. This remaining supersymmetry enabled him to apply his ideas about the relation between supersymmetric quantum theories and topology, and finally to get the right theory.

As Atiyah had wanted, this quantum field theory was such that for any given four-dimensional space, if one tried to use it to compute observable quantities, one got zero most of the time. The only situation in which one did not get zero was for certain quantities that were independent of deformations of the space. These were exactly Donaldson's new topological invariants, called the Donaldson polynomials. While Donaldson had defined these mathematical objects and proved that they had certain properties, for a general four-dimensional space they were extremely difficult actually to compute. Witten hoped that his quantum field theory would allow him to compute the Donaldson polynomials in many cases, but these hopes were initially not realized. If one applied the standard semiclassical technique of perturbation expansion about a solution of the classical field equations, one just got back one of the definitions of his polynomials already known to Donaldson. Mathematicians whose specialty was four-dimensional topology were not impressed by Witten's results. From their point of view, he had just given a more complicated definition of the Donaldson polynomials, in terms of a quantum field theory that there was little hope of defining rigorously.

Witten quickly moved on to apply the idea of topological quantum field theory to a number of other cases, generating a large class of new quantum field theories, each of which had topological invariants as their observable quantities. One of these cases turned out to be particularly interesting and surprising. A subfield of topology with a long history is that of the theory of knots. To a topologist, a knot is something like a piece of string that lies in a complicated pattern in three-dimensional space, with its two ends tied together. If one moves the string around, deforming it in the three-dimensional space, some knots can be untangled, while others can't. One of the central goals of knot theory is to find topological invariants one can associate with each knot. These are invariant when one deforms the knot, for instance while trying to untangle it. Ideally, one would like a topological invariant that would be the same as the invariant of the trivial untangled knot only if the knot in question could be untangled. Then for any knot, to see whether it could be untangled, all one would have to do is compute the invariant and see whether it was the same as the invariant of the untangled knot.

The Jones polynomial that Atiyah had mentioned at the Weyl conference was a topological invariant that was at the center of a lot of research by knot theorists. It had also made an appearance in some work on two-dimensional conformal quantum field theories. Spurred on by Atiyah, Witten tried to see how it could be fit into a quantum field theory, finally realizing how to do this during discussions over dinner with Atiyah and Atiyah's former student Graeme Segal at Annie's restaurant during a conference in Swansea in the summer of 1988. Ten years later a plaque was unveiled at the restaurant to commemorate the occasion. By September, Witten had produced a topological quantum field theory whose physical quantities were precisely the Jones polynomials. This quantum field theory with three space-time dimensions was a deceptively simple theory, having a gauge symmetry but no supersymmetry. It was built out of Yang–Mills gauge fields, and the knot appeared as the trajectory of an infinitely heavy charged particle moving in three-dimensional space-time. The equation for the so-called Lagrangian function that determines the dynamics of the theory had exactly one term in this case. This term is a subtle mathematical quantity built out of the Yang–Mills fields of the gauge theory and called a Chern–Simons term, after the geometers Shiing-Shen Chern and James Simons, who had first investigated it in 1971. Chern was one of the greatest geometers of the twentieth century, passing away in 2004 at the age of 93, while Simons, after building up an excellent mathematics department at Stony Brook, left academia to found Renaissance Technologies, one of the world's most successful hedge funds. Witten's new topological quantum field theory quickly became known as the Chern–Simons or Chern–Simons–Witten theory.

The Chern–Simons theory could be defined for any three-dimensional space, so it gave not only the Jones polynomials for knots in standard three-dimensional space, but analogues for any other space with three dimensions. The most surprising part of the theory was its Hilbert space. The Hilbert space was finite-dimensional with a dimension given by the Verlinde formula first discovered in conformal field theory. Packed into Witten's new quantum field theory defined by its single Chern–Simons term were amazing and unexpected relations among the topology of knots and three-dimensional spaces, the theory of Kac–Moody groups and their representations, conformal field theories, index theory, and much else besides.

At the International Congress of Mathematicians in Kyoto in 1990, Witten was awarded the Fields Medal, the most prestigious honor in mathematics, an award for which his work on the Chern–Simons theory was largely responsible. There is no Nobel Prize in mathematics, so the Fields Medal is the closest analogue (although the recently instituted Abel Prize may change that). It is a somewhat different sort of prize, typically given to from two to four mathematicians once every four years, unlike the Nobel Prize in physics, which is given each year to from one to three physicists. In addition, Fields medalists must be under the age of forty when they receive the award. The first Fields Medal was given in 1936, and no physicist had ever received it before Witten. It is unlikely that any physicist before Witten had even been considered. At the time of the award to Witten, it wasn't universally popular in the mathematics community. There was a feeling among many that since Witten was not working with rigorously precise definitions and was not giving rigorous proofs of theorems, what he was doing might be interesting, but it was not really mathematics. Some of the doubters were topologists who worked in four dimensions, and thus were familiar mainly with Witten's four-dimensional topological quantum field theory, which to them didn't seem to have anything new to say about the Donaldson polynomials.

Witten continued thinking off and on about how to get new information about Donaldson polynomials out of his topological quantum field theory. During the late 1980s and early 1990s, many mathematicians became interested in the mathematics of these topological invariants, and an active subfield of topology grew up that was known as Donaldson theory. Steady progress was made in this kind of mathematics, but the technical problems involved were substantial, so things moved slowly. This all changed very dramatically in the fall of 1994.

When one considers Maxwell's equations for just the electromagnetic field, ignoring electrically charged particles, one finds that the equations have some peculiar extra symmetries besides the well-known gauge symmetry and space-time symmetries. The extra symmetry comes about because one can interchange the roles of the electric and magnetic fields in the equations without changing their form. The electric and magnetic fields are said to be dual to each other, and this symmetry is called a duality symmetry. Once electric charges are put back in to get the full theory of electrodynamics, the duality symmetry is ruined. In 1931, Dirac realized that to recover the duality in the full theory, one needs to introduce magnetically charged particles with peculiar properties. These are called magnetic monopoles and can be thought of as topologically nontrivial configurations of the electromagnetic field, in which the electromagnetic field becomes infinitely large at a point. Whereas electric charges are weakly coupled to the electromagnetic field with coupling strength given by the fine-structure constant $\alpha = \frac{1}{137}$, the duality symmetry inverts this number, demanding that the coupling of the magnetic charge to the electromagnetic field be strong with strength $\frac{1}{\alpha} = 137$.

If magnetic monopoles exist, this strong coupling to the electromagnetic field would make them easy to detect. All experiments that have looked for them have turned up nothing (with the possible exception of an experiment at Stanford in 1982, which saw one candidate and nothing thereafter). During his first visit to Oxford in 1978, Witten had met the physicist David Olive, who, together with Claus Montonen, had conjectured that there might be an analogue of the electric–magnetic duality symmetry in the four-dimensional Yang–Mills case. Witten saw that this conjecture could be made most plausibly for a supersymmetric version of Yang–Mills theory, and soon published a paper with Olive on the topic. Over the years, Witten sometimes returned to this idea, and in the spring of 1994, together with Nathan Seiberg, he was able to work out an explicit solution of a supersymmetric Yang–Mills theory that had a version of the conjectured duality. This was a dramatic development, since it finally gave an example of a quantum field theory of Yang–Mills type in which one could understand explicitly what was happening at strong coupling. In addition to the standard perturbation expansion, which told one what was happening at weak coupling, in this theory one could calculate things at strong coupling using a dual picture to the one used in the weak-coupling calculation. This dual picture involved magnetic monopoles and gauge fields that were not Yang–Mills fields, but instead had the simpler U(1) gauge symmetry, the same as in QED.

Witten realized that this new solution also had something to say about the topological quantum field theory for the Donaldson polynomials, since the topological theory was, up to his twisting trick, essentially the same theory as the one he and Seiberg had solved. He went to Cambridge to give a talk about his work with Seiberg to the physicists at MIT on October 6, 1994. Noticing that several mathematicians were in the audience, at the end of the talk he mentioned that this work might be related to Donaldson theory and wrote down an equation that he thought should be the dual analogue of the ones mathematicians had been studying so far (the so-called self-duality equations). The mathematicians couldn't make much sense of most of his talk, but several of them started to think about the new equation that Witten had shown them.

Harvard's Clifford Taubes was one of the mathematicians in the audience; he soon saw the potential implications of the new equation Witten had just written down and immediately began studying it. Taubes couldn't see how to justify mathematically Witten's claim that the solutions of this new equation were related to the solutions of the self-duality equation, but he could see that the solutions of the new equation were a lot easier to understand. They involved U(1) gauge fields rather than the SU(2) Yang–Mills gauge fields used by Donaldson, and the fact that U(1) is a commutative group makes things much easier than the case of SU(2), which is noncommutative. Taubes soon realized that everything he and other experts in Donaldson theory had been working so hard to do with the self-duality equations could be done instead more or less trivially using Witten's new equation. One of Taubes's colleagues at Harvard, the mathematical physicist Arthur Jaffe, describes what happened next:

So after that physics seminar on October 6, some Harvard and MIT mathematicians who attended the lecture communicated the remark [about the new equation] by electronic mail to their friends in Oxford, in California, and in other places. Answers soon began to emerge at break-neck speed. Mathematicians in many different centers gained knowledge and lost sleep. They reproved Donaldson's major theorems and established new results almost every day and every night. As the work progressed, stories circulated about how young mathematicians, fearful of the collapse of their careers, would stay up night after night in order to announce their latest achievement electronically, perhaps an hour, or even a few minutes before some competing mathematician elsewhere. This was a race for priority, where sleep and sanity were sacrificed in order to try to keep on top of the deluge of results pouring in. Basically ten years of Donaldson theory were re-established, revised, and extended during the last three weeks of October 1994.[7]

Taubes gave a talk on November 2 at Harvard entitled "Witten's Magical Equation" and announced the death of Donaldson theory and the birth of a new field, which came to be called Seiberg–Witten theory, based on replacing the self-duality equations with the Seiberg–Witten equation. Witten had caused an entire subfield of mathematics to be revolutionized in a few short weeks, simply by telling the experts which equation to think about. Any topologists specializing in four-dimensional topology who had been at all skeptical about the value of Witten's mathematical ideas became instant converts, convinced that, at least ex post facto, he had richly earned his Fields Medal.

In addition to the Chern–Simons and Donaldson topological quantum field theories that led to new ideas about the topology of three- and four-dimensional spaces, as well as knots in three dimensions, Witten came up with a third sort of topological quantum field theory in early 1988. The existence of this third kind of theory was again something that Atiyah had conjectured in his talk at Duke the year before. Witten called this new quantum field theory a topological sigma model, in reference to the so-called sigma model used in current algebra. In general, physicists now call a quantum field theory a sigma model if its fields associate with each point in space-time not a number or vector, but a point in a target space, which is a curved space of some dimension. The sigma model that appears in current algebra is one in which the fields associate a group element with each point in space-time, so the target space is just a group. The space of all possible elements in a group is a curved space of some dimension. For the group U(1) it is just the circle, of dimension one. For the group SU(2) it is the three-dimensional sphere, the analogue of the two-dimensional surface of a sphere, in one dimension higher.

Witten's topological sigma model was a two-dimensional quantum field theory that was a sigma model whose target space had something called a complex structure. A space is said to have a complex structure if at every point on the space, nearby points can be labeled by coordinates that are complex numbers. Geometrically, this means that at each point a 90-degree rotation of the coordinates has been singled out, one that determines what happens when one multiplies coordinates by the square root of minus one. It turns out that not all spaces have complex structures. An obvious condition the space must satisfy is that its dimension must be an even number, since each complex coordinate corresponds to a pair of real coordinates.

In the topological sigma model both the two-dimensional space-time and the target space have complex structures, so one can impose an analyticity condition on fields, much like the one discussed earlier for the case of conformal transformations. This condition roughly says that a field is analytic if multiplying the coordinates of either space-time or the target space by the square root of minus one gives the same field. While in general there is an infinite number of possible field configurations, the number of them that are analytic is a lot smaller, sometimes even zero or a finite number.

The observable quantities in Witten's topological sigma model were essentially the numbers of these analytic field configurations. These numbers were the analogues in this model of the Donaldson polynomials in Witten's first topological quantum field theory. Now it turns out that the problem of computing such numbers is part of the field of mathematics known as algebraic geometry. Algebraic geometry is a branch of mathematics with a long and complicated history, and it reached a high degree of sophistication during the last half of the twentieth century. In simplest terms, algebraic geometry is the study of solutions to sets of polynomial equations. These sets of equations may have no solutions, a finite number of solutions, or an infinity of solutions. In high-school mathematics one learns how to find the roots of a polynomial of one variable (the solutions of the equation given by setting a polynomial to zero). Algebraic geometers study the analogous problem, but now for more than one polynomial and more than one variable.

As in the case of a single polynomial of a single variable, the whole subject simplifies if one uses complex variables, and this is what algebraic geometers often do. When a set of polynomial equations has an infinity of solutions, one can think of these solutions as being the points of a new space. Such spaces whose points are solutions to polynomial equations can be very nontrivial, and they are the main object of study of algebraic geometry. When the polynomial equations are equations in complex variables, these solution spaces can be given complex coordinates. These solution spaces are the kind of space that could be a target space for Witten's topological sigma model, so one might hope that this quantum field theory contains new information about them. The general idea is that for each solution space, the topological sigma model will give one a number (the number of analytic fields), and this number is a sort of topological invariant. Given two different solution spaces, one way of proving that they are really different is to compute the number of analytic fields in the two cases and show that these numbers are different.

Just as Witten's first topological quantum field theory did not actually tell topologists anything about Donaldson invariants that they did not already know, the topological sigma model also did not actually tell algebraic geometers anything about numbers of analytic fields that they did not already know. During the next few years, many physicists studied the model, and what they learned about it in the end made things very interesting for mathematicians. The topological sigma model was a supersymmetric quantum field theory, and Witten had used the same trick of twisting a supersymmetric theory that he had used in the Donaldson case. In addition, it was an example of a conformal field theory, since its observable quantities were invariant under all deformations of two-dimensional space-time, including the conformal transformations. Much was known about conformal field theories by 1988, and this information was put to work. One thing that was known about this type of supersymmetric conformal field theory was that given one such theory, there was a simple transformation one could do on it to get a new, different, but closely related one. If one did the transformation twice, one got back the original theory. This is a lot like the mirror-reflection symmetry transformation, so this was called a mirror symmetry.

If one started with a certain target space, constructed the corresponding topological sigma model (and thus a conformal field theory), then did the mirror reflection, what could one say about the new conformal field theory? Was it another topological sigma model, but for a different target space? If so, the new target space was called the mirror space of the original one. The physicists Brian Greene and Ronen Plesser found such a pair of mirror spaces in 1990, and Philip Candelas and his collaborators at the University of Texas studied many examples starting in 1991. One of the examples Candelas et al. found turned out to be of great interest to the algebraic geometers. It involved a target space known to them as the "quintic threefold," since it was the space of solutions of a fifth-degree polynomial and had three complex dimensions (thus six real dimensions). If one takes two-dimensional space-time to be the surface of a sphere, the problem of counting the number of analytic fields for the quintic threefold was well known to algebraic geometers. Analytic fields could be classified by an integer, the so-called degree, and the cases of degree one and two were already understood. The number of such fields of degree one was known since the nineteenth century to be 2875, and the number for degree two had been calculated to be 609,250 by Sheldon Katz in 1986. The calculation for degree three was in progress, but mathematicians didn't know how to go beyond that. It wasn't even known whether there was a finite number of analytic fields for each degree, although there was a conjecture of algebraic geometer Herbert Clemens that this was true.

The Candelas group was able to do something that stunned the mathematicians. By doing a calculation of a different kind on the mirror space, they were able to get a formula that gave the numbers of analytic fields for all degrees at once. This was the same sort of thing that was to happen in the Seiberg–Witten story: the original topological quantum field theory didn't make things easy to compute, but it could be related to another one, where the calculation was dramatically easier. At first, mathematicians were dubious about the result. The physicist's mirror-space method predicted that there were 317,206,375 analytic fields of degree three, but a calculation by two mathematicians had just given a different answer. Soon the mathematicians found an error in their calculation, and when it was done correctly, they found the same number as the physicists. This result impressed many algebraic geometers, whose field of mathematics traditionally had little or nothing to do with physics. Atiyah, one of whose mathematical specialties was algebraic geometry, was heard to remark that now he and his colleagues would have to read about new progress in their own field in the main journal of the particle theorists, *Nuclear Physics B*.

During the last decade, the field of mirror symmetry has been a very active one, with a continuing interchange between mathematicians and physicists, each with their own perspectives. Much of the effort of mathematicians has been devoted to trying to formulate in precise language and prove rigorously conjectures made by physicists using the language of quantum field theory. Some of the newer mathematical proofs exploit ideas about the symmetry groups of the spaces involved, a part of the story that made little appearance in the initial mirror-symmetry work. Physicists have been exploring a dizzying array of connections between topological sigma models (especially a variant known as the "topological string"), Witten's Chern–Simons version of gauge theory, "matrix models" involving integrals over groups such as $SU(N)$ for large N, and much else besides. A great deal of this work is motivated by attempts to understand large-N gauge theories in terms of string theory, in the simplified context in which both the gauge theory and the string theory are not the full physically interesting ones, but instead versions that carry just topological information. Many new conjectures involving unexpected relations between different areas of mathematics continue to emerge, opening up new and exciting problems for both mathematicians and physicists to explore.

FURTHER READING

Unfortunately, as far as I know, there are no nontechnical expositions available for the topics discussed in this chapter. Some expository but still quite technical sources that one could consult for these topics are as follows:

For instantons, see *Geometry of Yang–Mills Fields*,[8] by Atiyah, and Coleman's 1977 Erice lectures[9] on the uses of instantons.

For lattice gauge theory, see *Lattice Gauge Theories: An Introduction*,[10] by Rothe.

For anomalies, see the volume *Current Algebra and Anomalies*,[11] by Treiman, Jackiw, Zumino, and Witten, as well as two articles by Atiyah: "Anomalies and Index Theory"[12] and "Topological Aspects of Anomalies."[13]

For large N, see Coleman's 1979 Erice lectures[14] on $\frac{1}{N}$.

For two-dimensional and topological quantum field theories, see the two-volume set *Quantum Fields and Strings: A Course for Mathematicians*,[15] derived from lectures given at the Institute for Advanced Study, edited by Deligne et al., as well as *Mirror Symmetry*,[16] by Hori et al.

String Theory

History

Until this point this book has been largely an enthusiastic tale of scientific successes, and as promised, those ideas that have not led to much progress have been ruthlessly suppressed and ignored. In reality, the successful ideas described in detail here were often pursued by only a small minority of physicists, with the great majority of their colleagues following very different research programs that were ultimately to fail. Beginning with this chapter, attention will turn to the history of some of the ideas that haven't worked out and how they have affected theoretical physics up to the present day. Readers who like their science always to be inspirational are advised that now may be the time to stop reading this book and to find a version of this story told by someone with a much more positive point of view regarding it. Some tales of this kind can be found in Brian Greene's *The Elegant Universe*[1] and *The Fabric of the Cosmos*[2] as well as Michio Kaku's *Hyperspace,*[3] *Beyond Einstein: The Cosmic Quest for the Theory of the Universe,*[4] and *Parallel Worlds.*[5]

S-MATRIX THEORY

From the early days of quantum field theory during the 1930s, the initially discouraging situation with the problems caused by infinities led to many proposals for alternatives. Some of these proposals took the point of view that since the infinities came from interactions of the fields at very short distances, one should do away with the concept of fields defined at all points. The idea was that at short distances something else should replace the field, but no one was able to find something else that worked as well as quantum field theory.

Many of the early pioneers in quantum theory were heavily influenced by the Vienna school of logical positivism. One of the tenets of this philosophical program was that one should try to develop science in such a way that one never needed to refer to metaphysical objects, meaning objects that were not directly accessible to perception. This idea made a lot of sense to many physicists as they struggled with the new quantum theory, finding that quite a few classical concepts, such as that of a particle having a definite position and momentum, needed to be abandoned. Such classical concepts were identified as metaphysical, and a consistent viewpoint on quantum physics required trying to avoid thinking about them.

One such positivistic approach to particle theory originated with John Wheeler in 1937, and was further developed by Heisenberg in 1943. This became known as the S-matrix philosophy, since the idea was that one should express the theory purely in terms of the scattering matrix. The scattering matrix is the mathematical quantity that tells one what happens if one has two particles that are initially far apart and one sends them toward each other. Do they bounce off each other, emerging from the collision intact but moving in a different direction? Do they annihilate each other, producing other particles? The S-matrix answers these questions, which are precisely the ones experimentalists are equipped to study. A quantum field theory can be used to calculate the S-matrix, but it inherently contains the much more complicated structure of fields interacting with each other at every point in space and time. Unlike a quantum field theory, the S-matrix is something that has nothing to say about exactly what is going on as the two particles approach each other and their interaction evolves.

Pauli was highly skeptical of Heisenberg's S-matrix ideas, noting in 1946 that "Heisenberg did not give any law or rule which determines mathematically the S-matrix in the region where the usual theory fails because of the well-known divergencies. Hence his proposal is at present still an empty scheme."[6]

Pauli's point was that the S-matrix proposal did not actually solve any of the physical problems that had motivated it. While it might allow one to avoid talking about what was happening at short distances, which was thought to be where the problems of infinities originated, the problems were still there in the end results of one's calculations.

The success of renormalized QED in dealing with the infinities eliminated one motivation for the S-matrix philosophy, but it remained the most popular way of thinking about the strong interactions throughout the 1950s, 1960s, and early 1970s, up until the advent of QCD. It seemed clear to almost everyone that there was no way that quantum field theory could explain the increasingly large number of distinct strongly interacting particles. In the early 1960s, the leading figure in strong interaction theory was Geoffrey Chew, who, with many collaborators at Berkeley and elsewhere, pursued a version of S-matrix theory called the analytic S-matrix. Here "analytic" means that a special condition is imposed on the structure of the S-matrix, an analyticity condition on how the S-matrix varies as the initial energies and momenta of the incoming particles are varied. This condition is the same mathematical condition discussed earlier in various contexts and requires working with energies and momenta that take on complex values. This analyticity property of the S-matrix is reflected in certain equations called dispersion relations. Chew and others believed that together with a couple of other general principles, the analyticity condition might be enough to predict the S-matrix uniquely. By the end of the 1950s, Chew was calling this the bootstrap philosophy. Because of analyticity, each particle's interactions with all others would somehow determine its own basic properties, and instead of having fundamental particles, the whole theory would somehow "pull itself up by its own bootstraps."

By the mid-1960s, Chew was also characterizing the bootstrap idea as nuclear democracy: no particle was to be elementary, and all particles were to be thought of as composites of each other. This democracy was set up in opposition to the aristocracy of quantum field theory, in which there were elementary particles: those that corresponded to the quanta of the fields of the theory. In Berkeley in the mid-1960s one definitely didn't want to be defending aristocracy and denigrating democracy. By this time, the quark model was having considerable success in classifying the strongly interacting particles, and this posed a challenge to Chew's ideas, since it was based on a picture of quarks as fundamental and other particles as composites of them. In 1966, near the end of one of his books on S-matrix theory, Chew asked,

> In the absence of experimental evidence for strongly interacting aristocrats, why should there be resistance to the notions of a complete nuclear democracy? Put another way, why are quarks popular in certain circles?[7]

He partly answered his own question a little later:

> The third reason for dislike by some of a dynamically governed democratic structure for nuclear society, with no elementary particles, is that it makes life exceedingly difficult for physicists. We must await the invention of entirely new techniques of analysis before such a situation can be thoroughly comprehended,

which was one way of saying that the theory really wasn't working out as hoped. David Gross, then a student of Chew's at Berkeley, recalls attending a talk in 1966 during which he finally realized that the bootstrap program was "less of a theory than a tautology."[8]

In retrospect, S-matrix theory is nothing more than a characterization of some of the general properties that the S-matrix computed from a quantum field theory is going to have. As far as one can tell, there are many consistent quantum field theories, and in particular, many variations on QCD appear to be consistent, so there are many different strongly interacting theories, all with different S-matrices. The bootstrap program's hopes that there would somehow be a unique consistent S-matrix were simply wishful thinking.

Besides involving fundamental particles, the successes of the quark model were very much due to the exploitation of the mathematics of the SU(3) group of symmetries and its representations. The partisans of nuclear democracy were in a losing battle not only with elementary fields, but also with the idea that symmetry is a fundamental principle. The earlier quotation from Chew comes from a section of his aforementioned book called "Aristocracy or Democracy; Symmetries versus Dynamics," in which he attempts to argue that the relation of symmetries to fundamental laws is remote. Here the fundamental laws are the postulated properties of the S-matrix that govern the dynamics of the theory, and have nothing to do with groups or representations. The division into two camps pursuing symmetry and dynamics was also remarked on by Feynman, who, commenting on the S-matrix theorists' fondness for dispersion relations, supposedly quipped, "There are two types of particle theorists: those who form groups and those who disperse."[9]

The dominance of S-matrix theory was international, perhaps even stronger in the Soviet Union than in the "People's Republic of Berkeley." According to Gross, "In Berkeley as in the Soviet Union, S-matrix theory was supreme and a generation of young theorists was raised ignorant of field theory. Even in the calmer East Coast, S-matrix theory swept the field."[10]

One effect of this was visible even to a student trying learn quantum field theory during the mid-1970s. While there were many recent textbooks about S-matrix theory, the most recent textbook on quantum field theory was one written by James Bjorken and Sidney Drell a dozen years earlier, and it was very much out of date, since it did not cover Yang–Mills theory.

The S-matrix program continued to be pursued by Chew and others into the 1970s. Just as the political left in Berkeley fell apart, with many turning to Eastern and New Age religions, followers of the S-matrix also stopped talking about democracy, and some began to look to the East. The physicist Fritjof Capra received a PhD in 1966, working with Walter Thirring in Vienna, but by the early 1970s had turned to Eastern religion, finding there deep connections to S-matrix theory. His book *The Tao of Physics* was first published in 1975. It extensively contrasts Western notions of symmetry with what he sees as Eastern ideas about the dynamic interrelationship of all things.[11] For instance,

The discovery of symmetric patterns in the particle world has led many physicists to believe that these patterns reflect the fundamental laws of nature. During the past fifteen years, a great deal of effort has been devoted to the search for an ultimate "fundamental symmetry" that would incorporate all known particles and thus "explain" the structure of matter. This aim reflects a philosophical attitude which has been inherited from the ancient Greeks and cultivated throughout many centuries. Symmetry, together with geometry, played an important role in Greek science, philosophy and art, where it was identified with beauty, harmony and perfection . . .

The attitude of Eastern philosophy with regard to symmetry is in striking contrast to that of the ancient Greeks. Mystical traditions in the Far East frequently use symmetric patterns as symbols or as meditation devices, but the concept of symmetry does not seem to play any major role in their philosophy. Like geometry, it is thought to be a construct of the mind, rather than a property of nature, and thus of no fundamental importance . . .

It would seem, then, that the search for fundamental symmetries in particle physics is part of our Hellenic heritage which is, somehow, inconsistent with the general world view that begins to emerge from modern science. The emphasis on symmetry, however, is not the only aspect of particle physics. In contrast to the "static" symmetry approach, there has been a "dynamic" school of thought which does not regard the patterns as fundamental features of nature, but attempts to understand them as a consequence of the dynamic nature and essential interrelation of the subatomic world.[12]

Capra then went on to write two chapters explaining the inadequacy of quantum field theory and the wonders of the bootstrap philosophy. *The Tao of Physics* was completed in December 1974, and the implications of the November Revolution one month earlier that led to the dramatic confirmations of the standard-model quantum field theory clearly had not sunk in for Capra (like many others at that time). What is harder to understand is that the book has now gone through several editions, and in each of them Capra has left intact the now out-of-date physics, including new forewords and afterwords that with a straight face deny what has happened. The foreword to the second edition of 1983 claims, "It has been very gratifying for me that none of these recent developments has invalidated anything I wrote seven years ago. In fact, most of them were anticipated in the original edition,"[13] a statement far from any relation to the reality that in 1983 the standard model was nearly universally accepted in the physics community, and the bootstrap theory was a dead idea. The afterword includes truly bizarre and counterfactual statements such as, "QCD has not been very successful in describing the processes involving strongly interacting particles."[14] In the afterword to the third edition, written in 1991, Capra writes worshipfully of Chew:

[He] belongs to a different generation than Heisenberg and the other great founders of quantum physics, and I have no doubt that future historians of science will judge his contribution to twentieth-century physics as significant as theirs . . .

Chew has made the third evolutionary step in twentieth-century physics. His bootstrap theory of particles unifies quantum mechanics and relativity theory into a theory that represents a radical break with the entire Western approach to fundamental science.[15]

Even now, Capra's book, with its nutty denials of what has happened in particle theory, can be found selling well at every major bookstore. It has been joined by some other books on the same topic, most notably Gary Zukav's *The Dancing Wu-Li Masters*. The bootstrap philosophy, despite its complete failure as a physical theory, lives on as part of an embarrassing New Age cult, with its followers refusing to acknowledge what has happened.

THE FIRST STRING THEORIES

The bootstrap philosophy's main idea was the hope that the analyticity condition on the S-matrix, together with some other conditions, would be enough to determine it uniquely. The problem with this idea is that there is an infinity of S-matrices that satisfy the analyticity condition, so the other conditions are crucial, and no one could figure out what they should be. Calculations were performed using the perturbation series expansion of a quantum field theory to produce an S-matrix, which was then examined to see whether some of its properties could be abstracted as general conditions. This method never led to a consistent way of dealing with the theory outside of the context of the perturbation expansion.

In 1968, the physicist Gabriele Veneziano noticed that a mathematical function, first studied by the mathematician Leonhard Euler during the eighteenth century and called the beta function, had the right properties to describe an analytic S-matrix. This S-matrix was quite unlike the ones coming from perturbation expansions. It had a property called duality, which in this context meant that looking at it in two different ways told one about two different kinds of behavior of strongly interacting particles. This duality has nothing whatsoever to do with the duality between electric and magnetic fields discussed earlier.

From 1968 on, this dual S-matrix theory was all the rage, with a large proportion of the particle theory community working on it. By 1970, three physicists (Yoichiro Nambu, Leonard Susskind, and Holger Bech Nielsen) had found a simple physical interpretation of Veneziano's formula. They found that it could be thought of as the S-matrix for a quantum-mechanical theory that corresponded to a classical mechanical system in which the particles were replaced by strings. A string is meant to be a one-dimensional path in space, an idealization of the position occupied by a piece of string sitting in some configuration in three-dimensional space. Such strings can be open, meaning they have two ends, or closed, meaning the two ends are connected. Whereas it takes only three numbers to specify the position of a particle in space, specifying the position of a string takes an infinite collection of numbers, three for each point on the string.

Standard techniques of quantum mechanics could be applied to the problem of how to get a quantum theory for the string, and much was learned over the next few years about how to do this. This is a tricky problem, but the final result was that physicists were able to get a quantum theory of the string, but one that had two serious problems. The first was that the theory really worked only if the dimension of the space and time that the string lives in is twenty-six, not four. The second problem was that the theory included a tachyon. To a particle theorist, a tachyon is a particle that moves faster than the speed of light, and if such a thing occurs in a quantum field theory, it is an indication that the theory is going to be inconsistent. One problem is that tachyons can transmit information backward in time, thus violating the principle of causality. In a theory in which causality is violated, there is a danger that one can imagine going back in time and killing an ancestor, thus rendering one's very existence an inconsistency. Theories with tachyons also generally lack a stable vacuum state, since the vacuum can just decay into tachyons.

Another obvious problem with string theories was that they did not include any fermions. Recall that these are particles with half-integer spin, like the electron and proton. To make any contact with the real world of strong-interaction physics, this problem had to be solved. The first string theory with fermions was constructed by Pierre Ramond late in 1970. He did this by generalizing the Dirac equation from its well-known version with three space variables to the case of the infinite number of variables needed to describe the string. During the next few years, many physicists worked on string theories with fermions, and it was found that this kind of string theory could be made sense of in 10 dimensions rather than in the 26 dimensions of the original string. This was still not the correct four dimensions, but at least it was somewhat closer.

Further work showed that a version of supersymmetry was at work and was an important part of string theory with fermions. Recall that in 1971–1973, several groups had discovered the idea of supersymmetry for quantum field theories with four space-time dimensions, and that this new symmetry is some sort of square-root of translation symmetry involving fermions. If one looks at the surface swept out by a string as it moves, one gets a two-dimensional space called the world sheet of the string. One way of thinking of a string theory is in terms of two-dimensional quantum field theories defined on these world sheets. Early string theorists discovered that string theories with fermions involved a version of supersymmetry that is an analogue of the four-dimensional supersymmetry, but instead in the two dimensions of the world sheet. This was actually the impetus for one of the independent discoveries of four-dimensional supersymmetry, that of Wess and Zumino in 1973. This kind of string theory is now known as a superstring theory, although that terminology did not come into use until much later.

For a few years, this early superstring theory was the leading candidate for a theory of the strong interaction. Many physicists were very impressed by its properties, and by the fact that there was a new theory to investigate that was not a quantum field theory. Susskind reports that in the early 1970s, "David Gross told me that string theory could not be wrong because its beautiful mathematics could not be accidental."[16] While string theorists were increasingly worried that the theory seemed to be in strong disagreement with the experimental results on deep inelastic scattering from SLAC, string theory remained very popular until the discovery of asymptotic freedom in 1973. By mid-1973, the implications of the asymptotic freedom idea had started to sink in, and most physicists quickly abandoned work on string theory and shifted over to working on QCD.

One person who continued working on superstring theory was John Schwarz, a student of Chew's who arrived at Caltech in 1972. While others were abandoning the subject for QCD, Schwarz continued investigating the superstring, since he felt strongly that "string theory was too beautiful a mathematical structure to be completely irrelevant to nature."[17] One of the many problems that superstring theory had as a theory of the strong interactions was that it predicted the existence of an unobserved strongly interacting massless particle of spin two. In 1974, together with Joel Scherk, Schwarz proposed that this particle should be identified with the graviton, the quantum of the gravitational field. They conjectured that superstring theory could be used to create a unified theory that included both the Yang–Mills fields of the standard model and a quantum field theory of gravity. This idea wasn't very popular at the time, but over the next few years, Schwarz and a small number of others worked off and on trying to make sense of it. By 1977 it was shown that in superstring theory the vibrational modes of the superstring that were bosons could be matched up with those that were fermions, in the process getting rid of the longstanding problem of the tachyon. This also indicated that the superstring had not only a supersymmetry on the two-dimensional world sheet, but a separate ten-dimensional supersymmetry like that of four-dimensional supersymmetric quantum field theories.

In 1979, Schwarz began collaborative work on superstring theory with the British physicist Michael Green. Over the next few years, they made a great deal of progress in formulating an explicitly supersymmetric version of the theory and learning how to do calculations with it. During this period, Schwarz was still employed by Caltech, but in a nonfaculty position. The fact that he was working in such an unfashionable area as superstring theory meant that he was not considered a reasonable candidate for a tenured faculty position. Both the popularity of superstring theory and Schwarz's job prospects were soon to change dramatically.

THE FIRST SUPERSTRING THEORY REVOLUTION

By 1983, Witten had begun to take an increasing interest in superstring theory. In April, at the fourth in the sequence of conferences on grand unified theories held during the 1980s, he gave a general talk about the prospects for a unified theory based on superstrings.[18] This talk was just a survey of the work of others such as Schwarz and Green, and included no new results of his own. While Witten's interest in the subject was not widely known, one of his students, Ryan Rohm, was working on superstring theory. He published a paper on the subject that year.

Even though his interest in the theory was growing, there was one potential problem for the theory that Witten felt was quite serious. Recall that a gauge anomaly is a subtle effect caused by how quantum field theories are defined that can ruin the gauge symmetry of a theory. This then implies that the standard methods for making sense of the theory will no longer be valid. During 1983, Witten was much concerned with gauge anomalies and suspicious that superstring theory would be inconsistent because of them. In a paper on gauge anomalies published in October 1983, Witten noted that an example of a theory in which the gauge anomalies under consideration canceled was the low-energy limit of a version of superstring theory. There were several different versions of superstring theory, and the one with the gauge anomaly cancellation was called type II. In this version of superstring theory there was no way to incorporate the Yang–Mills fields of the standard model, but there was another version of superstring theory, called type I, in which this was possible. The question whether the type I theory also had a gauge anomaly was still open.

During the summer of 1984, Green and Schwarz, working together at the Aspen Center for Physics, a sort of summer camp for physicists, finally managed to calculate the anomalies in the type I theory. There are many different versions of type I superstring theory, and they found that while in almost all versions the theory had a gauge anomaly, there was one version in which the various gauge anomalies cancelled out. This happened for the version in which the symmetry group was the group SO(32), meaning rotations in real 32-dimensional space. Green and Schwarz knew from a phone conversation with Witten that he was very interested in this result, and so sent him a copy of their paper via FedEx (this was before e-mail attachments) at the same time they sent it off to the journal *Physics Letters B*. This was on September 10, and the flurry of activity it was to set off became known to string theorists later as the First Superstring Revolution. The real date for this revolution should, however, be perhaps eighteen days later on September 28, when the first paper on superstring theory by Witten arrived at the same journal as the Green–Schwarz paper. By itself, the news that gauge anomalies cancel in a version of type I superstring theory would probably not have had so dramatic an effect on the particle theory community, but the news that Witten was now devoting all his attention to this idea spread among theorists very quickly.

At Princeton, a group of four physicists (David Gross, Jeff Harvey, Emil Martinec, and Ryan Rohm) quickly found another example of a superstring theory in which gauge anomalies cancel. This theory was given the name heterotic superstring theory, using a term from genetics denoting a hybrid. The four physicists involved were later to be jokingly called the Princeton String Quartet. Their paper on the heterotic superstring arrived at the journal on November 21. Witten was part of another group of four physicists who rapidly worked out the details of a proposal for how to get the physics of the standard model from the heterotic superstring. Their paper arrived at the journal on January 2. During the next few years, a huge proportion of particle theorists started working on superstring theory. Many of them had worked on the early version of the theory that predated QCD, and so they just had to pick up where they had left off a decade earlier. SLAC maintains a very comprehensive database of the high-energy physics literature called SPIRES, which is indexed by keyword. It lists 16 papers on superstrings in 1983, 51 in 1984, 316 in 1985, and 639 in 1986.

By then, work on the superstring completely dominated the field, a situation that has continued to some degree to the present day.

Several factors account for this spectacularly quick shift in particle theory research. One is certainly that by 1984 there were few good untried ideas around to work on, so many physicists were looking for something new to do. Another is the fact that the superstring was not completely new to many physicists, since they had worked with string theory early in the previous decade. By far the most important factor was Witten, who was at the height of his influence in physics. He believed strongly in the theory, worked very hard at understanding it, and promoted it heavily. I know of more than one physicist who, during this period, went to talk to Witten in Princeton about their non-string-theory work, only to be told that while what they were doing was all well and good, they really should drop it and start working on superstring theory.

What was this heterotic superstring theory that caused all the excitement? Like all known superstring theories, it was a theory of strings moving in ten space-time dimensions. The variables describing the strings had an additional symmetry group acting on them consisting of two copies of something called E_8. The group E_8 is a Lie group much like $SU(2)$ and all the other ones that have found a use in particle theory, but it does have some special properties. Whereas the other Lie groups described earlier have a geometric interpretation as groups of rotations of vectors with real or complex coordinates, E_8 is one of five exceptional Lie groups that have no such interpretation.

E_8 is the largest of the five exceptional groups, and it corresponds to a 248-dimensional space of possible symmetry transformations. This high dimension and the lack of a geometric definition mean that calculations with E_8 must be performed by rather intricate algebraic methods of a specialized kind. The exceptional groups in general and E_8 in particular have a reputation among mathematicians as being quite obscure mathematical objects. The following rather peculiar quotation is an extract from a letter the topologist Frank Adams included at the end of one of his papers, a letter purporting to be written by E_8 (not the sort of thing one generally finds in math papers):

Gentlemen,

Mathematicians may be divided into two classes; those who know and love Lie groups, and those who do not. Among the latter, one may observe and regret the prevalence of the following opinions concerning the compact exceptional simple Lie group of rank 8 and dimension 248, commonly called E_8.

(1) That he is remote and unapproachable, so that those who desire to make his acquaintance are well advised to undertake an arduous course of preparation with E_6 and E_7 [two smaller exceptional Lie groups].

(2) That he is secretive; so that any useful fact about him is to be found, if at all, only at the end of a long, dark tunnel.

(3) That he holds world records for torsion [a subtle topological invariant].

. . .

[a refutation of these points follows]

. . .

Given at our palace, etc, etc,
and signed
E_8.[19]

E_8 is such a large group of symmetries that one can easily include the much smaller symmetry group SU(5) used in grand unified theories inside it. One can even include two larger groups often used in grand unified theories: SO(10) (rotations of 10-dimensional real vectors) and E_6 (another exceptional group). Thus in principle, one can hope to set up the heterotic string theory in such a way as to include a grand unified theory as its low-energy limit.

A trickier problem to deal with is the difference between the ten-dimensional space-time in which the superstring must be formulated and the four-dimensional space-time of the real world. One can hypothesize that for each point in our four-dimensional space-time there really is an unobservably small six-dimensional space, giving the universe a total of ten dimensions, with only four of them big enough to be seen. So perhaps the world is really ten-dimensional, with six of the dimensions too small to be observed. This is a version of the Kaluza–Klein idea that was discussed earlier in the context of eleven-dimensional supergravity. In that case there were seven inconvenient dimensions that needed to be explained away.

There are various consistency conditions one would like the theory to satisfy. A fundamental postulate is that the predictions of the superstring theory should not depend on conformal (angle-preserving) transformations of the two-dimensional string world sheet. Imposing this condition and requiring supersymmetry of the theory, one can show that the six-dimensional space must be one that can be described at each point in terms of three complex coordinates, and its curvature must satisfy a certain condition. This condition on the curvature is one that only certain six-dimensional spaces can satisfy. The mathematician Eugenio Calabi conjectured in 1957 that all that was needed for this curvature condition to be satisfiable was that a certain topological invariant had to vanish, and in 1977 this was proven by Shing-Tung Yau. Spaces that satisfy this condition on the curvature are now called Calabi–Yau spaces in their honor.

The predictions of the heterotic string theory strongly depend on which Calabi–Yau space one chooses. In 1984, only a few Calabi–Yau spaces were known, but by now hundreds of thousands have been constructed. It is not even known whether the number of Calabi–Yau spaces is finite or infinite, and two of my algebraic-geometer colleagues have a bet between them over how this will turn out. The English algebraic geometer Miles Reid states, "I believe that there are infinitely many families, but the contrary opinion is widespread, particularly among those with little experience of constructing surfaces of general type."[20]

Each of these potentially infinitely many Calabi–Yau spaces is actually a family of possible spaces, since each Calabi–Yau space comes with a large number of parameters that describe its size and shape.

Throughout the late 1980s and 1990s, much effort by physicists was devoted to the construction and classification of new sorts of Calabi–Yau spaces. This research led to significant interaction between physicists and mathematicians, the most important of which surrounded the issue of "mirror symmetry" discussed earlier. These years also saw a great deal of work on two-dimensional quantum field theories, especially conformally invariant ones, since these were what came up in the construction of superstring theories.

THE SECOND SUPERSTRING THEORY REVOLUTION

By the early 1990s, interest in superstring theory was beginning to slow down. There were five known consistent kinds of string theory:

- The SO(32) type I theory, whose anomaly cancellation was discovered in 1984.
- Two variants of type II superstring theory.
- The heterotic string theory with two copies of E_8 symmetry.
- A variant of the heterotic string theory with SO(32) symmetry.

For various technical reasons, it was thought that the E_8 heterotic string was the most promising theory to pursue, and most work was devoted to its study. In a talk at a string theory conference at the University of Southern California in March 1995, Witten unveiled a remarkable set of conjectures about how these five theories were interrelated. He described evidence that had been accumulated over the past few years that there were various duality relations between these five theories. He also gave evidence for a duality sort of relation between string theories and supergravity theory in eleven dimensions. Recall that supergravity theory is a quantum field theory based on a supersymmetric version of Einstein's general relativity. As a quantum field theory, it is known to have renormalizability problems, and eleven dimensions is the largest number of space-time dimensions in which it can be consistently constructed. Back in 1983, Witten had shown that attempts to use it as a unified theory, in which seven of the eleven dimensions were assumed to be small, had inherent problems reproducing the mirror-asymmetric nature of the weak interactions.

A crucial part of Witten's web of new conjectures was the conjectural existence of a new supersymmetric eleven-dimensional theory. This was to be a theory whose low-energy limit was supergravity, but at higher energies contained new things that were not describable by quantum fields. To have the right properties to satisfy his conjectures, it had to have not one-dimensional strings, but two- and five-dimensional "p-branes." Here p is some nonnegative integer, and a p-brane is a p-dimensional space that can move inside the eleven-dimensional space. A string is a 1-brane, and a 2-brane is a two-dimensional surface moving around in eleven-dimensional space. One can visualize such a two-dimensional surface as a membrane, and this is the origin of the "brane" terminology.

Ever since the early days of string theory, some physicists have investigated the possibility of membrane theories, which would be theories of fundamental objects of higher dimensionality than the one dimension of the string. Attempts to define theories of this kind by analogy with string theory have not been very successful, since they lead to technical problems that so far seem to be insoluble. Thus the 11-dimensional theory conjectured to exist by Witten cannot be constructed simply by using an analogue of string theory, and one must hope for a completely new kind of theory that for some unknown reason can describe 2-branes and 5-branes. Witten dubbed this conjectural theory M-theory, with the explanation that "M stands for magic, mystery or membrane, according to taste."[21] Since 1995, a great deal of effort has gone into trying to figure out what M-theory is, with little success. The most successful attempt uses infinite-dimensional matrices, and the name "matrix theory" provides yet another possible version of what the M is supposed to represent. The matrix theory formalism works only for certain special choices of the geometry of the eleven dimensions, and in particular, it does not work in the physically relevant case of four of the 11 dimensions being large and seven very small.

Witten's most grandiose conjecture of 1995 was that there is a single underlying theory that reduces in six different special limiting cases to the five known superstring theories and the unknown M-theory. This largely unknown general theory is also often referred to as M-theory, and yet another explanation for the M is "mother," as in "mother of all theories." Glashow jokes in a *Nova* television series on string theory[22] that the M is really an upside-down W for "Witten." In the same TV show Witten says, "Some cynics have occasionally suggested that M may also stand for 'murky,' because our level of understanding of the theory is, in fact, so primitive. Maybe I shouldn't have told you that one."

There is not even a proposal at the present time about what exactly this theory is. The formulation of the general conjecture of the existence of an M-theory and the more specific conjectures about dualities have led to a very large amount of work by many physicists. In the process, 1995 has sometimes become known as the date of the Second Superstring Theory Revolution. From this time on, the name "superstring theory" becomes something of a misnomer, since those working in this field now feel that they are studying bits of a much larger theory that contains not only strings but higher-dimensional p-branes. Besides M-theory, this theory has also been sometimes called the "theory formerly known as strings."

RECENT TRENDS

In November 1997 there appeared a paper by Juan Maldacena containing a new idea that has dominated recent research in string theory. This idea is variously referred to as the Maldacena conjecture or the AdS/CFT conjecture. This conjecture posits a duality relation between two very different kinds of theories in two different dimensions. One of the theories is a four-dimensional supersymmetric version of Yang–Mills quantum field theory, one with $N = 4$, i.e., four different supersymmetries. This theory has been known for a long time to be a rather special quantum field theory, since it has the property of being scale invariant. In other words, the theory has only massless particles and thus nothing to set a distance or energy scale. This scale invariance also implies a conformal invariance, meaning invariance under four-dimensional changes of coordinates that don't change angles. This conformal invariance makes the theory a conformal field theory, explaining the CFT part of the AdS/CFT conjecture name (note that this kind of four-dimensional conformal quantum field theory is quite different from the two-dimensional conformal field theories mentioned earlier).

The theory related by duality to this CFT is a theory of superstrings, now in a five-dimensional space of a special type. This space (or at least its analogue in four dimensions) is known to those who study the curved spaces of general relativity as anti-deSitter space. Willem de Sitter was a mathematician who studied this kind of space in the early twentieth century, but with an opposite sign for the curvature, whence the "anti." This use of anti-deSitter space explains the AdS part of the AdS/CFT acronym. Anti-deSitter space is a five-dimensional space of infinite size, and the AdS/CFT conjecture says that the theory of superstrings moving in it has a duality relation to the four-dimensional CFT described earlier. So the duality has the hard-to-understand feature of relating a string theory in five dimensions to a quantum field theory in four dimensions. The four dimensions are supposed to be in some sense the four-dimensional space of directions in which one can move infinitely far away from any given point in the anti-deSitter space. This type of duality is sometimes referred to as holographic: just as a hologram is a two-dimensional object encoding information about three dimensions, here a quantum field theory in four dimensions encodes information about what is happening in one more, namely five, dimensions.

Since it is about superstrings in five infinitely large dimensions, the AdS/CFT idea does not obviously help one learn anything about the case one cares about in which there are only four infinitely large dimensions. Those working on the conjecture hope that it can be generalized, in particular to the case in which the four-dimensional quantum field theory at issue is not the conformally invariant supersymmetric Yang–Mills theory, but perhaps QCD, the Yang–Mills theory with no supersymmetries. If this were the case, it would provide a specific realization of the longstanding hope that there might be some sort of string theory that is dual to the QCD quantum field theory. One then hopes that doing calculations in this dual theory would be possible and would finally provide a real understanding of the long-distance behavior of QCD.

The amount of work done in the past seven years on the AdS/CFT conjecture is truly remarkable. To date (spring 2006), the SLAC SPIRES database of papers on high-energy physics shows 3,944 articles that cite Maldacena's original article. In the history of particle physics there are only two more heavily cited papers, and in both of those cases the large number of citations come from the fact that the papers concern subjects about which there has been a huge amount of experimental activity. No other specific speculative idea about particle physics that has not yet been connected to anything in the real world has ever received anywhere near this amount of attention.

Since 1998 another very popular topic among theorists has been something called "brane-world scenarios." The original superstring or M-theory ideas for constructing a unified theory assume that of the ten or eleven dimensions required by the theory, four are the space-time dimensions we see, and the remaining six or seven dimensions are very small. In a brane-world scenario, some or all of these six or seven dimensions may be much larger than in the original picture, and some mechanism is assumed to exist that keeps the fields of the standard model confined to the observed four dimensions, preventing them from spreading out into the other dimensions. By appropriately choosing the sizes and properties of these extra dimensions, one can construct models in which there are observable effects at energy scales reachable by planned or conceivable accelerators.

In the last few years many string theorists have stopped working toward a better understanding of string theories and have moved into the field of cosmology, creating a new field called "string cosmology." The central issue in cosmology is the study of the physics of the very early universe, and astronomers have had great success in recent years performing new observations that shed light on this problem. String cosmologists hope that superstring theory can be used to make predictions about what happened at the very high energy scales that must have been in play in the very early universe.

Finally, the most recent trend in superstring theory revolves around the study of what is known as the landscape of a vast number of possible solutions to the theory. As noted in the first chapter of this book, a controversy now rages as to whether this sort of research is a complete abandonment of traditional notions of what it means to do theoretical science. This rather bizarre turn of events will be considered in detail in a later chapter.

The duality and M-theory conjectures of the second superstring revolution involve interesting issues about the geometry and topology of higher-dimensional spaces and have motivated some new ideas in mathematics. On the other hand, the brane-world scenarios, string cosmology, and landscape investigations that have played such a large role in particle theory during the past few years all mostly involve calculations that use only the very traditional mathematics of differential equations. The close and fruitful relationship between mathematics and theoretical physics that characterized much of the 1980s and early- to mid-1990s continues, but with lesser intensity as these latest subjects have come to dominate particle theory research.

FURTHER READING

For the history of supersymmetry, see *The Supersymmetric World: The Beginnings of the Theory*,[23] edited by Kane and Shifman.

For the history of superstrings, see the article "Superstrings—A Brief History," by John Schwarz.[24]

The standard textbooks on string theory are *Superstring Theory*,[25] by Green, Schwarz, and Witten; *String Theory*,[26] by Polchinski; and *A First Course in String Theory*,[27] by Zwiebach.

For branes and the AdS/CFT correspondence, see *D-Branes*,[28] by Johnson.

For a recent popular book on brane-world scenarios, see *Warped Passages*,[29] by Randall.

String Theory and Supersymmetry

An Evaluation

> There are repeated efforts with the symbols of string theory. The few mathematicians who could follow him might say his equations begin brilliantly and then decline, doomed by wishful thinking.
> —THOMAS HARRIS, *HANNIBAL*

As a general rule, scientific progress comes from a complex interaction of theoretical and experimental advances. This is certainly true of the standard model, although the sketchy way in which its history was told earlier in this book did not emphasize this point. In the course of the explanation of superstring theory and its history in the last chapter, the alert reader may have noticed the lack of any reference to experimental results. There is a good reason for this: superstring theory has had absolutely zero connection with experiment, since it makes absolutely no predictions.

This chapter will consider this peculiar state of affairs, and attempt to evaluate the progress that has been made in the last twenty years toward the goal of turning superstring theory into a real theory that can explain something about nature. Since the low-energy limit of superstring theory is supposed to be supersymmetric quantum field theory, this will begin with an examination of what is known about supersymmetric extensions of the standard model. This will be followed by an attempt to see what the problems are that keep superstring theory from really being a theory and what the prospects are for a change in this situation. It will end with a tentative attempt to evaluate the successes of supersymmetry and superstring theory in mathematics, where, in contrast to the situation in physics, there have been real accomplishments.

This chapter at times ventures into rather technical and detailed issues, and many readers may find some of it rough going. The reason for employing this level of detail is that these technicalities are unfortunately both important for what is being discussed and also not explained in most sources written for anything other than a very specialized audience. In any case, the hope is that the broad outlines of what is at issue here will remain clear to all readers.

SUPERSYMMETRY

Several conferences have been organized in recent years to celebrate the thirtieth anniversary of the idea of supersymmetry, and many of those involved in its development have recently written down their recollections of its early days.[1] At the time of writing, over 37,000 scientific papers on the subject of supersymmetry have been produced,[2] and they have continued to appear at an average rate of more than 1,500 a year for the past decade, a rate that shows no signs of decreasing. What has been learned, and what results are there to show for this unprecedented amount of work on a very speculative idea?

The idea of supersymmetry goes back to the early 1970s, and by the latter part of the decade an increasing number of researchers were working on the subject. According to the SLAC database, there were 322 papers on supersymmetry in 1979 and 446 in 1980. The subject then really took off, with 1,066 papers in 1982. Thus it was 1981 that saw a dramatic increase in the subject's popularity. In that year, Witten gave a series of lectures on the topic at the summer school for particle theorists held in Sicily, at Erice. This summer school had a long tradition as a venue at which some of the top particle theorists gathered with postdocs and graduate students to give survey lectures on the latest, hottest topics in the field. The organizer was a prominent Italian physicist, Antonio Zichichi, who was the subject of many legends, often involving his connections at the highest and lowest levels of Italian society. One story that made the rounds about Zichichi (I cannot recall who first told me this and have no idea whether it is true) was that one year, a physicist who was to lecture there had the misfortune of having his luggage stolen from him during the train trip to Erice. When he got there, he told Zichichi about the problem, remarking that it would be hard for him to give his lectures, since his lecture notes had been in his bags. Zichichi told him not to worry, that he was sure everything would be all right. The next morning he awoke to find that his luggage was waiting for him outside the door to his room.

Throughout much of the 1960s and 1970s some of the most famous lectures at Erice were those given by Harvard physicist Sidney Coleman, who lectured at the summer school approximately every other summer. The written versions of his magnificently lucid talks covering the ideas behind SU(3) symmetry, current algebra, asymptotic freedom, spontaneous symmetry breaking, instantons, the large-N expansion, and other topics were avidly read by just about every theorist as they appeared. They were collected in the book *Aspects of Symmetry*, published in the mid-1980s.[3] In 1981, Witten took over Coleman's role with a beautiful series of expository lectures on supersymmetry, including a general argument that continues to be considered one of the two main reasons for pursuing the idea.

Recall that from the mid-1970s on, a very popular idea for extending the standard model was that of grand unification, which involved constructing a Yang–Mills quantum field theory based on a symmetry group such as SU(5) or SO(10), one that was larger than the standard model SU(3) × SU(2) × U(1). Witten's argument was that any attempt to extend the standard model to a grand unified theory faced something called a "hierarchy problem." This means that the theory has a hierarchy of two very different energy (or equivalently, distance) scales, and it is very hard to keep them separate. The first is the energy scale of spontaneous symmetry breaking in the electroweak theory, which is responsible for the mass of the W and Z particles and is about 100 GeV. The second is the energy scale of the spontaneous symmetry breaking of the larger symmetry of the grand unified theory, and to avoid conflict with experiment, this must be at least 10^{15} GeV. Witten argued that if one introduced elementary fields (the Higgs fields) to accomplish this symmetry breaking of the vacuum state, then there was no natural way to ensure that one mass scale would be 10^{13} times smaller than the other. Unless one carefully "fine-tuned" to great accuracy every term in the perturbation expansion, the lower-energy scale would not stay small, and would end up of roughly the same size as the grand unification energy.

He then argued that supersymmetry provided a way out of this problem. It turns out that while there is no natural way of keeping masses of particles associated with bosonic fields like the Higgs field small, fermions that are not symmetric under mirror-reflection have a chiral symmetry that naturally keeps their masses zero. In a supersymmetric theory, bosons and fermions come in equal-mass pairs, so Witten's proposal was that the electroweak Higgs field was part of a supersymmetric theory and paired with a fermion whose mass could naturally be set to zero. Even in a theory with a large grand unification mass scale, the combination of supersymmetry and chiral symmetry could naturally keep the mass of the electroweak Higgs particle small. In summary, Witten's argument was that if one wanted a grand unified theory of the SU(5) type, and one wanted vacuum symmetry breaking done by a Higgs field, one had a problem (the hierarchy problem), and perhaps supersymmetry could solve it.

This argument was and continues to be highly influential, but one should keep in mind that it makes two crucial assumptions. The first is that there is grand unification, at a very high energy scale, with spontaneous symmetry breaking by a new set of Higgs fields. The second is that the mechanism for electroweak spontaneous symmetry breaking is an elementary Higgs field. Either or both of these assumptions could very well be wrong.

Besides Witten's hierarchy argument, there is a second argument for supersymmetry that has grown in influence over the last twenty years. This one also is based on the grand unification assumption. The grand unified theory is supposed to have only one number that characterizes the interaction strength, whereas the standard model has three, one for each of SU(3), SU(2), and U(1). One lesson of asymptotic freedom is that the interaction strength depends on the distance scale at which it is measured, so when considering grand unification, one needs to extrapolate the observed interaction strengths from accelerator energies where they are observed up to the grand unification energy scale. During the mid-1970s, when this calculation was first performed, it was observed that the three interaction strengths reached roughly the same value when they were extrapolated to about 10^{15} GeV, so this was assumed to be the grand unification scale. By now, there exist much more accurate measurements of the three observed interaction strengths. When one performs the same calculation, one finds that the three numbers don't quite come together at the same point. The three extrapolated energies at which one of the three possible pairs of interaction strengths coincide are all different, in the range 10^{13}–10^{16} GeV.

If one takes the most popular way of extending the standard model to a supersymmetric quantum field theory and redoes the same calculation, things improve considerably, and one gets the three interaction strengths coming together quite closely, at around 2×10^{16} GeV. To assign any significance to this result, however, one needs to make at least one very large assumption. This is that there is no new unknown physics taking place in the huge range of energy scales between what has already been studied (up to 100–1000 GeV) and the 2×10^{16} GeV scale. Such an assumption is known as the desert hypothesis. Since one doesn't know what the mechanism is that breaks the grand unified symmetry, another implicit assumption is that the three interaction strengths being equal at a given energy is actually a necessary part of the picture.

Besides these two arguments, some other kinds of arguments have been made for supersymmetric quantum field theories. Historically, an important motivation has been the fact that supersymmetry relates fermions and bosons, and thus one could hope that it would unify the known particles of the two types. Unfortunately, it is now well understood that this simply doesn't work at all, and such an argument is completely invalid. Given all of the known particles, any attempt to relate any two of them by supersymmetry is seriously inconsistent with experimental facts. The fundamental source of the problem is that supersymmetry is a space-time symmetry, independent of the internal SU(3) × SU(2) × U(1) symmetries. As a result, it must relate bosons and fermions that are in the same SU(3) × SU(2) × U(1) representations. Pairs of this kind do not occur in the standard model. This mismatch between the symmetry patterns that super-symmetry predicts and those that are observed continues to be true even for most hypothesized grand unified models, in which the extra particles needed for grand unification cannot be related to each other by supersymmetry.

Another argument often made for supersymmetry is that it is possible that the low-energy limit of a superstring theory is a supersymmetric quantum field theory. This argument is of course based on the assumption that at high energies the world is governed by a superstring theory. The next section will consider the arguments for that possibility, but one thing to keep in mind is that one of the main arguments often given for superstring theory is that it explains supersymmetry. These arguments are circular, and should best be interpreted as an argument that the fates of supersymmetry and string theory are linked, that these ideas are either both wrong or both right.

One other possible argument for the supersymmetry hypothesis would be that it leads to a compelling new theory that generalizes the standard model in a convincing way. A great deal of effort over the last twenty-five years has gone into the project of understanding all possible supersymmetric theories of this kind. The simplest supersymmetric theory generalizing the standard model goes under the name "minimal supersymmetric standard model," or MSSM, and will be considered next in some detail. Other viable possibilities must include the MSSM as just one part of the theory.

There are two fundamental problems that make it difficult to construct a simple supersymmetric extension of the standard model. The first is that, as mentioned earlier, since there is no way to use the supersymmetry to relate any known particles, it must relate each known particle to an unknown particle. For each known particle one must posit a new, unobserved particle called a superpartner, and by now there is even a detailed naming scheme for these particles. To each quark there is a squark, for each lepton a slepton, for each gluon a gluino, and so on. In addition, for the bosonic Higgs field one can't just associate a superpartner, but one needs to postulate a second set of Higgs fields with a second set of superpartners. If one doesn't double the number of Higgs fields, the theory will have an anomaly and some quarks cannot acquire nonzero masses.

The second fundamental problem is that these new particles one has postulated cannot have the same masses as the particles one already knows about, since otherwise, they would already have been observed. To stay consistent with experiments, one must assume that all of these new particles are so heavy that they could not have been produced and observed by current accelerators. This means that supersymmetry must be a spontaneously broken symmetry, because if it were a symmetry of the vacuum state, then one could show that each particle would have to have the same mass as its superpartner.

The necessity for spontaneous breaking of supersymmetry is a disaster for the whole supersymmetric quantum field theory project. Supersymmetric extensions of the standard model are well enough understood that it is clear that their dynamics are such that they cannot by themselves dynamically break their own supersymmetry, and if ones tries to break it in a similar way to the Higgs mechanism, one gets relations between particle masses that are incorrect. One can come up with ways of spontaneously breaking the supersymmetry, but these all involve conjecturing a vast array of new particles and new forces, on top of the new ones that come from supersymmetry itself. One of the best ways of doing this involves starting with a completely new hidden supersymmetric theory, one different enough from the standard model that it can break its own supersymmetry. The particles and forces of this new "hidden" theory have nothing to do with the known particles and forces, so now one has two completely separate supersymmetric quantum field theories. One then typically assumes the existence of a third "messenger" theory with its own set of particles, in which the particles of the third theory are subject to both the known forces and hidden forces. This third kind of particle is called a messenger particle, and there are various proposals for what kind of theory could govern it.

This whole setup is highly baroque and not very plausible, and furthermore, it completely destroys the ability of the theory to predict anything. Supersymmetry advocates describe this situation with phrases such as "the lack of a compelling mechanism for supersymmetry breaking." Since one doesn't understand the supersymmetry breaking, to define the MSSM one must include not only an unobserved superpartner for each known particle, but also all possible terms that could arise from any kind of supersymmetry breaking. The end result is that the MSSM has at least 105 extra undetermined parameters that were not in the standard model. Instead of helping to understand some of the 18 experimentally known but theoretically unexplained numbers of the standard model, the use of supersymmetry has added in 105 more. As a result, the MSSM is virtually incapable of making any predictions. In principle, the 105 extra numbers could take on any values whatsoever, and in particular, there is no way to predict what the masses of any of the unobserved superpartners will be. One can try to impose a condition on the theory that makes it more predictive by demanding that the energy scale of supersymmetry breaking be not too large (so that supersymmetry in some sense solves the hierarchy problem, which was one of the main motivations of the idea), but it is not at all clear in this case what "too large" means.

While the supersymmetry-breaking problems are by far the most damaging, the MSSM has some other undesirable features. To keep the superpartners from interacting with known particles in a way that disagrees with experiment, the theory must be set up in such a way that it has an R-parity symmetry. This is another mirror-reflection sort of symmetry, now arranged so that known particles stay the same under reflection while their superpartners do not. Even with this somewhat ad hoc constraint on the theory, there are still many ways in which it is in danger of disagreeing with experiment. The problem is that the standard model is simple enough that there are several classes of phenomena that can't happen within it, phenomena that have been looked for carefully and shown not to happen to a high degree of precision. In the MSSM these classes of phenomena are allowed, and will occur at rates in dramatic disagreement with experiment unless many of the 105 parameters are chosen to have very special values, something for which there is no known justification. The sorts of phenomena that generically occur in the MSSM but whose absence is predicted by the standard model include the following:

- Flavor-changing neutral currents. These are processes in which a quark changes flavor without changing its charge.

One of the main reasons the existence of the charmed quark was proposed in 1970 is that it automatically canceled out processes of this kind that would occur if it weren't there. The problem solved by the existence of the charmed quark now reappears in the MSSM.

- Processes in which one kind of lepton changes into another kind. An example is a muon decaying into an electron and a photon. This is energetically allowed, but is never seen in the real world.

- Large CP violation. Recall that the weak interactions are not invariant under the mirror-reflection or parity transformation. This transformation is conventionally denoted by P. The transformation that takes particles to antiparticles and vice versa is denoted by C. Thus CP is the transformation corresponding to doing C and P in succession. All experimental evidence is that physics is invariant under the CP transformation, except for one small effect that comes about once one has three or more generations of quarks. In the MSSM there are many potentially large sources of violation of this CP symmetry.

Another potential problem of the MSSM is called the "μ problem." This name refers to the conventional use of the Greek letter mu to denote the coefficient of the term in the MSSM that governs the mass of the supersymmetric Higgs particle. This problem is basically the danger of reemergence of the hierarchy problem that supersymmetry was supposed to solve. For everything to work, μ must be on the order of the electroweak spontaneous symmetry breaking energy scale, but there is no reason known for it not to be a factor of 10^{13} or more larger and at the grand unification scale. The mechanism used to solve the first hierarchy problem does not apply here, so one must do something else. One is forced to assume that for some unknown reason, whatever is happening at the grand unification scale doesn't make μ nonzero, and invoke some other fields not in the MSSM to give it a reasonable value.

Additional problems arise if one tries seriously to incorporate the MSSM in a grand unified theory. Recall that the simplest SU(5) nonsupersymmetric grand unified theory has now been ruled out by experiments that looked carefully for evidence of protons decaying into other particles and did not see this happening. Versions of SU(5) and other grand unified theories that are supersymmetric and have the MSSM as low-energy limit can be constructed, and the whole motivation of the MSSM is to allow this. Changing to a supersymmetric grand unified theory increases the grand unification scale somewhat, making the processes that caused protons to decay in the nonsupersymmetric version occur much more rarely. This gets the supersymmetric theory out of this particular trouble with experiment, but other processes can now occur that make it recur. An especially dangerous problem for the theory is something called the doublet–triplet splitting problem. The supersymmetric grand unified theory must contain not just the usual Higgs doublets, which are in the two-dimensional representation of the standard model SU(2), but also Higgs triplets, which are in the three-dimensional representation of the standard model SU(3). That is, they come in three colors, just like quarks. The mass of the Higgs doublet must be small, and the problems in the MSSM that come from trying to arrange this have been discussed earlier. On the other hand, the mass of the Higgs triplets must be very large, at the grand unified scale or higher, or they will cause protons to decay. Arranging for this large mass difference in a natural way is the doublet–triplet splitting problem. If one doesn't somehow arrange this, the theory will fail, since it will predict that protons decay at an observable rate.

To summarize the story so far, supersymmetry has exactly two features that can help one construct a grand unified theory. The first is that it allows one naturally to keep the mass scale of electroweak symmetry breaking and grand unification separate without fine-tuning the parameters of the theory. The second is that the strengths of the SU(3), SU(2), and U(1) standard-model forces extrapolate in a supersymmetric theory to be the same at around 2×10^{16} GeV. A crucial question is whether these two features add up to an experimental prediction. Supersymmetry inherently predicts that for every known particle there will be one we have not seen before, its superpartner. The first positive feature of the theory, the solution to the hierarchy problem, predicts that the masses of the superpartners can't be too different from the electroweak spontaneous symmetry breaking scale of about 200 GeV. For each of the superpartners, experiments rule out much of the mass range below this scale, but it is still possible that future experiments will see them if their masses are higher than this, but still not so high as to wreck convincingly the solution of the hierarchy problem. The second positive feature of supersymmetry can be read as making exactly one prediction: if the strengths of the three kinds of forces are to come together at precisely the same point, knowing two of them predicts the third. For instance, knowing the strength of the SU(2) and U(1) forces, we can predict the strength of the SU(3) strong force. This number is believed to be known experimentally to about 3 percent accuracy, and the prediction comes out 10–15 percent high.[4] It's hard to quantify the inaccuracy of this prediction precisely, since due to the way the mathematics works out, in doing the calculation in another way (turning it around to calculate the relative strengths of the other two forces in terms of the strength of the strong force), one gets a much more accurate prediction.

Thus the only fairly precise prediction of supersymmetry (broken at low energy) is off by 10–15 percent and requires assuming no new relevant physics all the way up to the GUT energy scale. The only less-precise predictions are that there should be many new particles in a general energy range, a significant part of which has been explored without finding them. The cost of these predictions is very high. In return for the not exceptionally accurate prediction of the value of one of its eighteen parameters, the standard model has to be replaced by a vastly more complex supersymmetric model with at least 105 new unexplained parameters. The new more complex model no longer explains many regularities of particle behavior that were explained in the standard model, while reintroducing the flavor-changing neutral current problem, among others.

A possible justification for taking seriously the MSSM despite its problems would be an aesthetic one. Perhaps the theory is just so beautiful that one can't help but believe that there must be something right about it. To make an informed judgment about this, one really needs to know the language of quantum field theory and see exactly how the MSSM is expressed in that language. Suffice it to say that I know of no one trained in quantum field theory who has ever characterized this particular theory as a beautiful one.

There is another motivation for considering supersymmetric field theories that has not yet been addressed. Recall that there are supergravity theories that are supersymmetric and also include gravitational forces. These theories have renormalizability problems, but one may have reason to believe that such problems can somehow be overcome. If one extends the MSSM not only to a supersymmetric grand unified theory, but even further to a theory that includes supergravity, then in principle one has a theory that describes all known forces, something every physicist very much desires. Unfortunately, this idea leads to a spectacular disagreement with observation.

The problem is yet again caused by the necessity of spontaneous supersymmetry breaking. It turns out that the quantity that measures whether supersymmetry is a symmetry of the vacuum state is exactly the energy of the vacuum. If the vacuum state is not invariant under supersymmetry, it will have nonzero energy. Recall that since we don't see equal-mass pairs of particles and their superpartners, the supersymmetry must be spontaneously broken. This means that the vacuum state must be noninvariant under supersymmetry and have a nonzero energy. The scale of this energy should be approximately the scale at which supersymmetry is spontaneously broken, which, we have seen, is at least a couple of hundred GeV. In supersymmetric grand unified theories, the vacuum energy will be much higher, since it will receive contributions from the grand unified energy scale.

The standard way one measures energy is as the energy difference with respect to the vacuum, and until one starts thinking about gravity, the energy of the vacuum is something that can never be measured, and can just be ignored. In Einstein's theory of gravity, general relativity, things are very different. The energy of the vacuum directly affects the curvature of space-time and occurs as a term in Einstein's equations that he called the cosmological constant. He initially put this term in his equations because he found that if it were absent, they predicted an expanding universe. Once astronomical observations showed that the universe is indeed expanding, the term could be set to zero and ignored. Over the years, astronomers have tried to make precise enough measurements of the expansion rate of the universe to see whether things indeed agree with the exact vanishing of the cosmological constant. In recent years, observations made of supernovae in very distant galaxies have indicated for the first time that the cosmological constant appears to be nonzero.

The value of the cosmological constant can be thought of as the energy density of the vacuum, or equivalently, the energy in a unit volume of space-time. Using units such that all energies are measured in electron volts (eV) and distances in inverse electron volts (eV^{-1}), the cosmological constant has units of eV^4, and astronomers believe its value is of order 10^{-12} eV^4. In a supersymmetric theory, spontaneous symmetry breaking must occur at an energy scale of at least 100 GeV = 10^{11} eV, which leads to an expected vacuum energy density of around $(100 \text{ GeV})^4 = 10^{44}$ eV^4. So the hypothesis of supersymmetry leads to an energy density prediction that is off by a factor of $10^{44}/10^{-12} = 10^{56}$. This is almost surely the worst prediction ever made by a physical theory that anyone has taken seriously. Supersymmetric grand unified theories make the situation much worse, since in them one expects contributions to the vacuum energy of order $(2 \times 10^{16} \text{ GeV})^4 = 1.6 \times 10^{101}$ eV^4, which is off by a factor of 10^{113}.

The problem with the cosmological constant is not a subtle one, and it is well known to everyone who works in the field of supersymmetry. When theorists publicly mention the problems with supersymmetry, this one is so basic that it is generally first on everyone's list. Many attempts have been made to find ways to find a way around the problem, but none has been successful. The most popular one in recent years involves essentially throwing up one's hands and deciding that the only way to understand the value of the cosmological constant is by anthropic reasoning. If it were considerably larger, the universe would expand too fast and galaxies wouldn't form. This does not really explain anything, but it has become a very popular argument nonetheless. As far as anyone can tell, the idea of supersymmetry contains a fundamental incompatibility between observations of particle masses, which require spontaneous supersymmetry breaking to be large, and observations of gravity, which require it to be small or nonexistent.

SUPERSTRING THEORY

Since physicists continue to take seriously the idea of a supersymmetric extension of the standard model, they must have a reason to believe that it may be possible to overcome the severe difficulties explained in detail in the last section. The most popular hope of this kind is that superstring theory will do the trick. This hope has motivated an unprecedented quantity of work by a very large number of the most prominent theoretical physicists for more than twenty years, but after all this time and effort, the whole project remains nothing more than a hope. Not a single experimental prediction has been made, nor are there any prospects for this situation to change soon.

The lack of any predictions of the theory makes many physicists very dubious that the idea is correct. One prominent theorist who felt this way up until his death in 1988 was Richard Feynman, who was quoted in an interview in 1987 as follows:

> Now I know that other old men have been very foolish in saying things like this, and, therefore, I would be very foolish to say this is nonsense. I am going to be very foolish, because I do feel strongly that this is nonsense! I can't help it, even though I know the danger in such a point of view. So perhaps I could entertain future historians by saying I think all this superstring stuff is crazy and is in the wrong direction.
>
> *What is it you don't like about it?*
>
> I don't like that they're not calculating anything. I don't like that they don't check their ideas. I don't like that for anything that disagrees with an experiment, they cook up an explanation—a fix-up to say, "Well, it still might be true." For example, the theory requires ten dimensions. Well, maybe there's a way of wrapping up six of the dimensions. Yes, that's possible mathematically, but why not seven? When they write their equation, the equation should decide how many of these things get wrapped up, not the desire to agree with experiment. In other words, there's no reason whatsoever in superstring theory that it isn't eight of the ten dimensions that get wrapped up and that the result is only two dimensions, which would be com-

pletely in disagreement with experience. So the fact that it might disagree with experience is very tenuous, it doesn't produce anything; it has to be excused most of the time. It doesn't look right.[5]

A more concise quotation I have heard attributed to Feynman on this topic is, "String theorists don't make predictions, they make excuses."[6]

Another Nobel Prize winner who has publicly attacked superstring theory on very similar grounds is Sheldon Glashow, who writes,

> But superstring physicists have not yet shown that their theory really works. They cannot demonstrate that the standard theory is a logical outcome of string theory. They cannot even be sure that their formalism includes a description of such things as protons and electrons. And they have not yet made even one teeny-tiny experimental prediction. Worst of all, superstring theory does not follow as a logical consequence of some appealing set of hypotheses about nature. Why, you may ask, do the string theorists insist that space is nine dimensional? Simply because string theory doesn't make sense in any other kind of space . . .
>
> Until the string people can interpret perceived properties of the real world they simply are not doing physics. Should they be paid by universities and be permitted to pervert impressionable students? Will young Ph.D.s, whose expertise is limited to superstring theory, be employable if, and when, the string snaps? Are string thoughts more appropriate to departments of mathematics, or even to schools of divinity, than to physics departments? How many angels can dance on the head of a pin? How many dimensions are there in a compactified manifold, 30 powers of ten smaller than a pinhead?[7]

The fundamental reason that superstring theory makes no predictions is that it isn't really a theory, but rather a set of reasons for hoping that a theory exists. At the first talk he gave on superstring theory, in Philadelphia in 1983,[8] Witten noted, "What is really unsatisfactory at the moment about the string theory is that it isn't yet a theory," and this remains true to this day. The latest name for the hopes that a theory may exist is M-theory, but to quote one expert on the subject, "M-theory is a misnomer. It is not a theory, but rather a collection of facts and arguments which suggest the existence of a theory."[9]

Yet another Nobel Prize winner, Gerard 't Hooft, while having kinder things to say about superstring theory than Feynman or Glashow, explains the situation as follows:

> Actually, I would not even be prepared to call string theory a "theory" rather a "model" or not even that: just a hunch. After all, a theory should come together with instructions on how to deal with it to identify the things one wishes to describe, in our case the elementary particles, and one should, at least in principle, be able to formulate the rules for calculating the properties of these particles, and how to make new predictions for them. Imagine that I give you a chair, while explaining that the legs are still missing, and that the seat, back and armrest will perhaps be delivered soon; whatever I did give you, can I still call it a chair?[10]

Why isn't superstring theory, which has been studied since the early 1970s, really a theory? To understand the problem, recall the earlier discussion of the perturbation expansion for QED, its problems with infinities and how they were solved by the theory of renormalization after World War II. Given any quantum field theory, one can construct its perturbation expansion, and (if the theory can be renormalized) for anything we want to calculate, this expansion will give us an infinite sequence of terms. Each of these terms has a graphical representation called a Feynman diagram, and these diagrams get more and more complicated as one goes to terms of higher and higher order in the perturbation expansion. There will be some parameter, or "coupling constant," that is typically related to the strength of the interactions, and each time we go to the next-higher order in the expansion, the terms pick up an extra factor of the coupling constant. For the expansion to be at all useful, the terms must get smaller and smaller fast enough as one goes to higher orders in the calculation. If this happens, one may be able to compute a few terms and then ignore all the higher ones, since they are small. Whether this happens will depend on the value of the coupling constant. If it is large, each higher-order term will get larger, and the whole thing will be useless. If it is small enough, each higher-order term should get smaller.

The best situation would be if the expansion is what is called a "convergent expansion." In this case, as one adds up terms of higher and higher orders, one gets closer and closer to the finite number that is the answer to one's problem. Unfortunately, it seems to be the case that this is not what happens for the renormalized perturbation expansions of nontrivial quantum field theories in four space-time dimensions. Instead, the expansion is at best an "asymptotic expansion," which means that two things are true. First the bad news: if one tries to add up all the terms, one will get not the right answer, but infinity. The good news, though, is that if one adds up only the first few terms, one can get something quite close to the right answer, and furthermore, one will get closer and closer to the right answer for smaller and smaller values of the coupling constant. This is what is going on in QED, where calculating low orders of Feynman diagrams gives results fantastically close to experiment, but where we have reason to believe that if one ever could calculate the terms with orders of several hundred or more, one would see that at higher orders the calculation would get worse instead of better. Since Yang–Mills quantum field theory is asymptotically free, the perturbation expansion is supposed to be an asymptotic expansion that is useful at short distances, getting better and better as one goes to shorter distances. At longer distances, it becomes useless because the effective coupling constant becomes large, and this is why different calculational techniques are needed to understand QCD fully. While the perturbation expansion method breaks down, Yang–Mills quantum field theory is a perfectly well defined theory, since one can rigorously define it using the lattice methods mentioned in an earlier chapter.

The situation in superstring theory is that for any physical process, what the theory offers is a method for assigning numbers to possible two-dimensional world-sheets swept out by moving strings. These world sheets can be organized topologically by counting the number of holes each one has. A superstring theory calculation gives one a number for zero holes, another number for one hole, yet another for

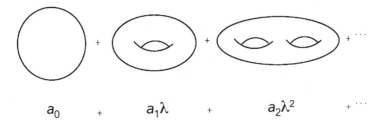

$$a_0 \quad + \quad a_1\lambda \quad + \quad a_2\lambda^2 \quad + \cdots$$

Series expansion in the number of holes.

two holes, and so on. The fundamental conjecture of superstring theory is that this infinite sequence of numbers is some kind of perturbation expansion for some unknown and well-defined theory. This conjectural underlying theory used to be referred to as nonperturbative superstring theory, but now is really what is meant by the term M-theory. No one knows what it is.

One might hope that expansion in the number of holes is actually a convergent expansion and thus good enough to calculate anything one wants to know to arbitrary precision, or that at least it is a good asymptotic series. There are strong arguments that the expansion is not convergent and that as one calculates more and more terms one gets a result that becomes infinite. In principle, the expansion could be a useful asymptotic series, but this is not quite what superstring theorists actually want to happen. There are some features of the terms in the superstring hole expansion that they like, and they want the expansion to be good for calculating those. There are other features that are very problematic, and they would like the expansion to fail completely when one tries to use it to calculate them.

The main features of the calculation that superstring theorists would like to keep are ones showing that in the low-energy limit, the theory looks like a theory of Yang–Mills fields and gravitons, since this was the main original motivation for superstring theory. There are quite a few features of the calculation that they would like to disown as things that should disappear in the true underlying M-theory. One of these features is the supersymmetry of the theory's vacuum. In the hole expansion, each term is exactly supersymmetric, with no spontaneous supersymmetry breaking. Somehow, whatever M-theory is, it is supposed to contain an explanation of where spontaneous supersymmetry breaking comes from. It should also allow calculation from first principles of the 105 extra parameters of the minimal supersymmetric standard model. Finally, it should solve all the problems of supersymmetric quantum field theories explained in the last section. There is no evidence at all for the existence of an M-theory that actually does this.

The other feature of the hole expansion that superstring theorists would like M-theory to get rid of is the so-called vacuum degeneracy. Recall that superstring theory makes sense only in ten dimensions. In the hole expansion there is an infinite list of possible ten-dimensional spaces in which the superstring could be moving. Superstring theory is a background-dependent theory, meaning that to define it one has to choose a ten-dimensional space as background in which the superstring moves. There is an infinite number of consistent choices of how to do this, only some of which have four large space-time dimensions and six small dimensions wrapped up in a Calabi–Yau space. There may or may not be an infinite number of possible Calabi–Yau spaces that would work, but no matter what, if one chooses one, it will be characterized by a large number of parameters that govern its size and shape. The vacuum degeneracy problem is that any Calabi–Yau space of any size and shape is equally good as far as the superstring hole expansion is concerned. In recent years, possible mechanisms have been found for fixing these sizes and shapes by adding some new structures into the problem, but these lead to a vast number of possibilities, and the implications of this will be examined in a later chapter. What superstring theorists would like M-theory to do is to somehow pick out a Calabi–Yau space of a specific size and shape, but again there is no evidence at all for the existence of an M-theory that does this.

There is one proposal for what M-theory is that has been investigated to some degree, the proposal called matrix theory ("matrix," as mentioned in the previous chapter, being one of many explanations for the M in M-theory). This proposal hasn't been made to work in cases one would like, but in the cases in which it does make sense, one finds that it does not solve the vacuum degeneracy problem.

There remain an infinite number of equally good solutions to the theory, with no way to choose among them.

When superstring theorists try to explain why superstring theory doesn't make any predictions, they often fall back on two explanations that are rather disingenuous. The first is that "solving the mathematics of the theory is just too difficult." As we have seen, that is not really the problem. The problem is that no one knows what equations to solve. It is true that higher-order calculations in the hole expansion of superstring theory are difficult, but one reason few physicists try to do them is that they know enough about the answer to know that it will have no supersymmetry breaking and will continue to have the vacuum degeneracy problem, so no predictions will be possible. The second explanation often used is that the basic energy scale of superstring theory is very high, so that characteristic superstring phenomena can never be observed, and extrapolations to what happens at low energy are difficult. This may be true, but the fact of the matter is that since there is no real theory, even if a particle accelerator were available that could reach these very high energies, superstring theorists would not be able to make any detailed predictions about what it would see. There are very general things one can say about what superstring theories should predict that quantum field theories do not, so some qualitative statements are possible about very high energies, but this is very different from having a real theory that makes real predictions.

Another standard argument for superstring theory is that it can predict things like the dimensionality of space-time (10) and the grand unification group (e.g., $E_8 \times E_8$) by insisting on anomaly cancellation. The problem with this is that these predictions are wrong, and to make them come out right requires the arbitrary choice of a Calabi–Yau space or something similar, ruining the predictive value of the theory. Since Witten's M-theory conjecture in 1995, superstring theorists are fond of saying that M-theory is unique, a claim whose significance is hard to evaluate, since one doesn't know what the theory is.

Superstring theorists often argue for the theory on the grounds that it will lead to a supersymmetric grand unified theory. We discussed such theories earlier and saw that they suffer from very serious problems and make no predictions, so this is not a very convincing motivation.

Various nonscientific arguments are frequently made for continuing to pursue research in superstring theory. The most common is that "it's the only game in town." The implications of this argument will be addressed in detail in a later chapter. In private, many physicists argue that Witten's strong support for superstring theory is in and of itself a very good reason to work in this area. Witten's genius and accomplishments are undeniable, and I feel that this is far and away the best argument in superstring theory's favor, but it is a good idea to keep in mind the story of an earlier genius who held the same position as Witten at the Institute for Advanced Study.

After Einstein's dramatic success with general relativity in 1915, he devoted most of the rest of his career to a fruitless attempt to unify electromagnetism and gravity using the sorts of geometric techniques that had worked in the case of general relativity. We now can see that this research program was seriously misguided, because Einstein was ignoring the lessons of quantum mechanics. To understand electromagnetism fully one must deal with quantum field theory and QED in one way or another, and Einstein steadfastly refused to do this, continuing to believe that a theory of classical fields could somehow be made to do everything. Einstein chose to ignore quantum mechanics despite its great successes, hoping that it could somehow be made to go away. If Witten had been in Einstein's place, I doubt that he would have made this mistake, since he is someone who has always remained very involved in whatever lines of research are popular in the rest of the theoretical community. On the other hand, this example does show that genius is no protection against making the mistake of devoting decades of one's life to an idea that has no chance of success.

Far and away the most common argument made for superstring theory is some version of "it's the only known consistent quantum theory of gravity." Stated this way, this statement is seriously misleading, although it is an attempt to refer to something with a stronger foundation. As we have seen, superstring theory is not really a theory, just a set of rules for generating what one hopes is the perturbation expansion of a theory. There is a sense in which this expansion is a significant improvement over the perturbation expansion one gets when one tries to treat general relativity by the standard methods of quantum field theory. As previously mentioned, applying the standard methods for generating perturbation expansions to the case of quantum gravity leads to a perturbation expansion that cannot be renormalized. If one computes higher-order terms in this expansion, for each new term one gets infinities that one doesn't know how to deal with.

This situation is improved in superstring theory, and this improvement is what the "only consistent quantum theory of gravity" argument is referring to. The calculation of higher-order terms in superstring theory is quite difficult, but there are some reasons to believe that the problems that make the higher-order terms in quantum field theory infinite may not be there in superstring theory. In brief, infinities in quantum field theory come from the short-distance behavior of the theory, and are related to the fact that interactions between two fields occur at precisely the same point in space-time. This is not how string theory works, so this source of infinities is not a problem. There are other sources of infinity one must worry about. For instance, what happens when the string becomes infinitely small? Recently, one of my colleagues at Columbia, the mathematician D. H. Phong, and his collaborator Eric D'Hoker (who was a student of David Gross and my contemporary at Princeton) have been able to understand precisely the structure of the two-hole term in the superstring hole expansion and to show that it does not have infinities. This required an impressively complex calculation, and higher-order terms should be even more difficult to understand. So the state of affairs now is that the 0-hole, 1-hole, and 2-hole terms have been proved to be finite, and the hope is that the higher-order ones are also problem-free.

This conjecture that superstring theory gives finite numbers for each term in the expansion is what leads some physicists to say that it is a consistent theory of gravity, but this ignores the fact that this is not a convergent expansion. While all the terms in the expansion may be finite, trying to add them all together is almost certain to give an infinite result. Actually, it would still be a problem for superstring theory even if the expansion were convergent. In that case, superstring theorists would have not just a perfectly consistent theory, but an infinity of them, all with features radically in disagreement with experiment (exact supersymmetry, degenerate vacuum states, and associated massless particles).

The other problematic aspect of the "only consistent theory" statement is the "only." There are various other proposals that have been made over the years for different ways to reconcile quantum mechanics and general relativity, but for none of these proposals has there been anything like the exhaustive investigation that has gone into superstring theory. One other proposal has attracted a sizable group of researchers, a proposal that goes under various names, one of which is "loop quantum gravity." In loop quantum gravity, the basic idea is to use the standard methods of quantum theory, but to change the choice of fundamental variables that one is working with. It is well known among mathematicians that an alternative to thinking about geometry in terms of curvature fields at each point in a space is to instead think about the holonomy around loops in the space. The idea is that in a curved space, for any path that starts out somewhere and comes back to the same point (a loop), one can imagine moving along the path while carrying a set of vectors, and always keeping the new vectors parallel to older ones as one moves along. When one gets back to where one started and compares the vectors one has been carrying with the ones at the starting point, they will in general be related by a rotation transformation. This rotation transformation is called the holonomy of the loop. It can be calculated for any loop, so the holonomy of a curved space is an assignment of rotations to all loops in the space.

Physicists working on loop quantum gravity have been making progress in recent years toward a consistent theory of quantum gravity, although it remains to be seen whether their framework is able to reproduce general relativity in the low-energy limit.[11] They take strong exception to the claims of superstring theorists to have the only quantum theory of gravity, which they find deeply offensive. For a popular account of loop quantum gravity, together with much more information about quantum gravity in general, one can consult Lee Smolin's recent book *Three Roads to Quantum Gravity*.[12] Perhaps the main reason the loop quantum gravity program has not attracted as much attention as superstring theory is that it is inherently much less ambitious. It is purely an attempt to construct a quantum theory of gravity and does not address the issue of how to unify gravity with the standard model. In particular, it does not claim to have a potential explanation of the eighteen parameters of the standard model.

Superstring theorists have another reason for believing that superstring theory can give a consistent quantum theory of gravity, a reason that has to do with calculations about black holes. Stephen Hawking was the first to show that if one combines quantum field theory with general relativity, one should find that black holes are not truly black, but that in fact, they emit radiation. The radiation is emitted as if the black holes were objects obeying the laws of thermodynamics with a temperature proportional to their area. Without a real quantum theory of gravity it has never been possible to check exactly how Hawking radiation works in a completely consistent theory. For certain special space-time backgrounds that one can interpret as limiting cases of black holes, superstring theorists have been able to show that Hawking radiation occurs as predicted. While these calculations cannot be performed for realistic black holes in four space-time dimensions, they are evidence of the existence of a consistent quantum theory of gravity being part of superstring theory.

A possibility consistent with everything known about superstring theory and loop quantum gravity is that just as there are many consistent quantum field theories that don't include gravity, there are also many consistent quantum theories—some field theories, some not—that do include gravitational forces. If the loop quantum gravity program is successful, it should construct a quantum theory of the gravitational field to which one could add just about any other consistent quantum field theory for other fields. If there is a consistent M-theory, it quite likely will depend on a choice of background space-time and make sense for an infinity of such choices. Neither loop quantum gravity nor M-theory offers any evidence for the existence of a unique unified theory of gravity and other interactions. Even if these theories do achieve their goal of finding a consistent quantum theory of gravity, if they don't have anything to say about the standard model, such theories will be highly unsatisfactory, since there is a serious question whether they can ever be experimentally tested. Distinctive quantum-gravitational effects occur at such high energy scales that it is very difficult to see how one can ever measure them. Perhaps some quantum-gravitational effect at the time of the Big Bang will somehow have an observable effect on cosmological models, and so be testable that way, but this remains unclear.

A final argument often heard for superstring theory is that the theory is simply so beautiful that it must somehow be true. This argument raises a host of issues (including that of whether the theory really is beautiful at all) and will be considered in detail in the next chapter. As long as no one quite knows exactly what string theory is, its proponents are able to hold very optimistic views about it. A statement often attributed to Witten (who credits the Italian physicist Daniele Amati) is that "superstring theory is a piece of twenty-first-century physics that fell by chance in the twentieth century." Some other statements of a similar kind are that superstring theory is a "supercomputer" or "spaceship" from the future, but one for which the instruction manual is lacking. These sorts of expressions function to some degree as an excuse for why superstring theorists are having trouble extracting predictions out of the theory, and as such, the first of them has become a lot less popular since the turn of the century. They also concisely express some other sentiments about the theory: that it is a baffling mystery, but one invested with great hopes, albeit ones that may never be fulfilled. What if the mysterious gadget that one hopes is a spaceship turns out to be merely a toaster?

Much of the appeal of superstring theory is thus not anything inherent in what is known about the actual theory itself, but rather a reflection of the hopes and dreams of the theorists who have devoted years of their lives to its study. In 1958, long before the standard model, Heisenberg worked with Pauli on a quantum field theory that they hoped would turn out to be a unified theory of the strong, electromagnetic, and weak interactions. They soon ran into problems with it, and Pauli became disillusioned, since they couldn't actually calculate anything. Heisenberg, on the other hand, remained convinced that they had made a discovery of great beauty and significance, and he gave public lectures and interviews to the press about their work. Pauli was greatly distressed by this and wrote letters to his colleagues, grumbling about Heisenberg's "radio advertisements" and dissociating himself from the whole thing. To some of them he included a drawing of a blank square with the notation, "This is to show the world that I can paint like Titian. Only technical details are missing."

Pauli's skepticism turned out to be justified, since Heisenberg's unified theory never led anywhere. Our modern understanding is that since the theory is nonrenormalizable, it is inherently incapable of ever reliably calculating anything. In a peculiar twist of history,[13] Volkov's original work on supersymmetry was motivated by an incorrect argument that the neutrino might be a Nambu–Goldstone particle, an argument contained in Heisenberg's book[14] about his unified field theory.

Some physicists have begun to come to the conclusion that superstring theory is much like Heisenberg's unified theory, a blank square aspiring to be a canvas by Titian, and that it will always be incapable of predicting anything about the real world. One of these is cosmologist Lawrence Krauss, who calls superstring theory a "Theory of Nothing."[15] Another is Daniel Friedan, the son of feminist Betty Friedan and one of the founders of perhaps the most prominent string theory group in the United States, at Rutgers University. Friedan, who received a MacArthur foundation "genius" grant in 1987 for his work related to string theory, recently wrote,

> String theory failed as a theory of physics because of the existence of a manifold of possible background spacetimes . . . The long-standing crisis of string theory is its complete failure to explain or predict any large distance physics. String theory cannot say anything definite about large distance physics. String theory is incapable of determining the dimension, geometry, particle spectrum and coupling constants of macroscopic spacetime. String theory cannot give any definite explanations of existing knowledge of the real world and cannot make any definite predictions. The reliability of string theory cannot be evaluated, much less established. String theory has no credibility as a candidate theory of physics.[16]

While Friedan's may still be a minority view among those who have worked on superstring theory, an increasing number of them are abandoning research aimed at trying to find out what M-theory is and working instead on things like brane-worlds and string cosmology as described in an earlier chapter of this book. Part of the motivation for this is a desire to respond to increasingly vocal critics within the physics community who have accused them of dealing in mathematics rather than physics. Another part of the motivation is that there simply are no promising ideas around concerning what M-theory might be.

Many other superstring theorists have abandoned work on the whole idea of using superstring theory to unify gravity and the standard model and instead are concentrating their attention in a much more promising area, that of trying to find a superstring theory dual to QCD. The AdS/CFT correspondence described earlier provides some hope that progress can be made in this direction, and also seems to offer possible new insight into superstring theory itself. Those working in this area sometimes make the point that these ideas may finally provide a nonperturbative formulation of at least one sort of superstring theory. Less often mentioned is that this conjectured nonperturbative definition of superstring theory consists in defining the theory as being precisely a certain quantum field theory. The lesson here for superstring theorists may very well be that quantum field theory is not something from which they can escape.

STRING THEORY, SUPERSYMMETRY, AND MATHEMATICS

While supersymmetry and string theory have been remarkably unsuccessful so far in explaining anything about physics, they have led to a great deal of new and very healthy interaction between the fields of mathematics and physics. In some sense it is this very lack of success as physics that has been responsible for much of the interesting interaction with mathematics, since it has led physicists to try out a wide range of ideas for overcoming the problems of these theories. If the initial hopes of 1984 had worked out and it had been possible to derive the standard model from string theory on a specific Calabi–Yau space, the interaction of mathematics and physics around this topic would have been much more limited in scope. Instead, physicists have found themselves investigating a mathematically very rich set of ideas, continually moving on to new ones as they find that the ones they already know about are insufficient to give them what they want.

Witten's new insights into a range of problems in quantum field theory during the 1980s brought a number of different mathematical ideas to the attention of physicists. The work of Witten and others using Calabi–Yau spaces to try to build realistic unified models using superstring theory was responsible for yet another wave of new mathematics entering physics. All of this activity changed the minds of many physicists who had previously been skeptical of the usefulness of abstract mathematics in physics. An example of this is Murray Gell-Mann, who despite his experience learning about the importance of SU(3) representation theory during the early 1960s had remained dubious about mathematics in general. By 1986, however, he was a convert, as can be seen from the following extract from his closing talk at a conference in that year:

> The importance of advanced mathematics in dealing with all the questions we have discussed is stupefying. Theoretical physics has been reunited with pure mathematics during the past decade in a most dramatic way, after an estrangement of half a century or so; and fundamental theoretical physics has been rejoined with the core of pure mathematics, where geometry, analysis and algebra (even number theory) come together . . .
>
> I hope that the trend in mathematical teaching, writing, and editing will continue to recoil from the extreme of Bourbakisme, so that explanations and nontrivial examples can be presented and physicists (to say nothing of other scientists) can once more have a fighting chance of understanding what mathematicians are up to, as they did early in the twentieth century.
>
> My attitude toward pure mathematics has undergone a great change. I no longer regard it as merely a game with rules made up by the mathematicians and with rewards going to those who make up the rules with the richest apparent consequences. Despite the fact that many mathematicians spurn the connection with Nature (which

led me in the past to say that mathematics bore the same sort of relation to science that masturbation does to sex), they are in fact investigating a real science of their own, with an elusive definition, but one that somehow concerns the rules for all possible systems or structures that Nature might employ . . . [17]

Gell-Mann's mention of "Bourbakisme" refers to the activities of a very influential group of French mathematicians known collectively by the pseudonym Bourbaki. Bourbaki was founded by André Weil and others during the 1930s, partly as a project to write a series of textbooks that would provide a completely rigorous exposition of fundamental mathematical results. They felt such a series was needed in order to have a source of completely clear definitions and theorems to use as a basis for future mathematical progress. This project came to fruition, and a steady stream of Bourbaki-authored volumes began to appear during the 1950s. These books are not very good places to try to learn mathematics, since they generally lack both examples and motivating exposition, concentrating on absolute clarity of definitions and complete rigor of argument. During this same period much research mathematics was aimed at trying to find the most general and most abstract formulation of mathematical results, and Bourbaki meant to provide a foundation for this work.

This kind of activity is what appalled Gell-Mann, and it did nothing for improving communication between mathematicians and physicists. While their books were arid and free of any examples, in their own research and private communications the mathematicians of Bourbaki were very much engaged with examples, nonrigorous argument, and conjecture. Besides writing the textbooks, their other main activity was the organization of the Séminaire Bourbaki. This was (and still is) held three times a year in Paris, and the format is that a half-dozen mathematicians are each delegated to prepare a one-hour talk and written report on a recent advance in mathematics, all done in a language as understandable to as wide a mathematical audience as possible. Bourbaki still exists, but the books have fallen by the wayside while the Séminaire continues.

The Bourbaki books and the point of view from which they emerged had a bad effect on mathematical exposition in general, with many people writing very hard-to-read papers in a style emulating that of the books. This trend began to dissipate by the 1970s, with a general turn in mathematics away from the use of abstraction for its own sake and more toward the exploration of specific examples. The influx of new ideas from physics during the 1970s, 1980s, and early 1990s accelerated this movement as it opened up several new areas for mathematicians to explore. These days, mathematicians generally have a more balanced attitude, often referring half-jokingly to some of the more abstract notions they use as "abstract nonsense" (as in "one can prove that particular theorem just by using abstract nonsense").

Gell-Mann's analogy of mathematics to masturbation, with physics instead the real thing, is a very well known one among physicists. There seems to be an issue of priority here, since (at least in its many appearances on the Internet) this analogy is generally attributed to Feynman. The successful use of sophisticated mathematics to gain better understanding of superstring theory during the mid- to late-1980s led many theorists, like Gell-Mann, to begin to change their attitudes toward mathematics. They began to think that perhaps mathematicians were on to something and not just engaged in intellectual onanism. Since the mid-1990s, the problems with superstring theory have led to something of a backlash against the use of mathematics in particle theory. Some theorists have laid superstring theory's failure ever to get close to making a real prediction at the door of abstract mathematics, rather than blaming any inherent problem with the underlying physical ideas. This backlash is reflected in the fact that most of the recent popular work on hot topics such as brane-worlds and string cosmology uses only very simple mathematics.

In his recent book *Faster Than the Speed of Light*, the cosmologist João Magueijo ends an extensive criticism of superstring theory and M-theory with this:

> To add to its mystique, the cult leader who coined the term never explained what the M stood for, and M-theorists heatedly debate this important issue. M for mother? M for membrane? M for masturbation seems so much more befitting to me.[18]

Elsewhere in his book Magueijo makes clear that he sees the root cause of the problems of superstring theory and M-theory to be an overly great attachment to mathematical elegance. He traces this fault back to Einstein:

> Unfortunately, Einstein himself bears a lot of the responsibility for having brought about this state of affairs in fundamental physics . . . He became more mystical and started to believe that mathematical beauty alone, rather than experimentation, could point scientists in the right direction. Regretfully, when he discovered general relativity—employing this strategy—he succeeded! And this experience spoiled him for the rest of his life . . . [19]

Instead of sneering at the supposedly masturbatory activities of mathematicians, physicists should perhaps consider whether George Orwell's rather nasty remark about political thinking among leftists might not equally well apply to much of modern particle theory: "Political thought, especially on the left, is a sort of masturbation fantasy in which the world of fact hardly matters." The question of who is having successful and satisfying intercourse with the deepest levels of reality, and who is just imagining it, still remains to be answered.

In recent years, mathematicians have been engaged in a slow process of integrating what they have learned from physicists into the main body of mathematical knowledge. Much effort is devoted to figuring out how to get precise statements of conjectures made by physicists and to find proofs of these statements, at least in limited cases. Since mathematically rigorous versions of quantum field theory do not exist, and many arguments in string theory do not even come from a well-defined physical picture, mathematicians can't just appropriate the physicists' arguments and try to make them rigorous. Once they have found a precise version of the purely mathematical implications of a conjecture coming from physics, they try to find a proof using known rigorous mathematical methods. This means that they are rarely able to get at the full content of the original idea as formulated in the language of quantum field theory. On the other hand, they are often getting ideas and conjectures about mathematical objects of a striking nature, since they come from a completely different conceptual framework from that in which these mathematical objects had originally been developed.

Many mathematicians are often little aware of the exact source in physics of the conjectures they study, whether it is quantum field theory, superstring theory, or M-theory. While superstring theory is a method for constructing terms in an expansion in a number of holes, each term in this expansion is a two-dimensional quantum field theory calculation. As a result, many calculations that are thought of as coming from superstring theory are actually results from two-dimensional quantum field theory. The calculations that are distinctly superstring theory calculations are those that involve not just one term at a time, but rather the structure of the entire set of terms in the hole expansion. One result of this lack of clarity in the mathematical community about what is a quantum field theory result and what is a superstring theory result has been that mathematicians are sometimes overly impressed by the whole idea of superstring theory.

On Beauty and Difficulty

But when the formula is very complex, that which conforms to it passes for irregular. Thus we may say that in whatever manner God might have created the world, it would always have been regular and in a certain order. God, however, has chosen the most perfect, that is to say the one which is at the same time the simplest in hypotheses and the richest in phenomena, as might be the case with a geometric line, whose construction was easy, but whose properties and effects were extremely remarkable and of great significance. I use these comparisons to picture a certain imperfect resemblance to the divine wisdom, and to point out that which may at least raise our minds to conceive in some sort what cannot otherwise be expressed. I do not pretend at all to explain thus the great mystery upon which depends the whole universe.

—G. W. LEIBNIZ, *DISCOURSE ON METAPHYSICS*

The late-seventeenth-century philosopher and mathematician Gottfried Wilhelm von Leibniz is widely known for his claim that we live in "the best of all possible worlds," a claim for which he has been mocked by authors from Voltaire[1] in the eighteenth century to Terry Southern[2] in the twentieth. A quick look around is enough to make one dubious about Leibniz's idea that a benevolent deity has chosen among all the possible worlds that one containing the minimal amount of evil, but Leibniz also had something more plausible in mind. Together with Isaac Newton, he was responsible for the discovery of differential and integral calculus, mathematical tools that in Newton's hands were used to formulate the extremely simple and powerful theory now known as Newtonian or classical mechanics. With some elaborations and the addition of fields as well as particles, classical mechanics was the foundational structure behind all of physics until special relativity and quantum mechanics came on the scene in the first decade of the last century. The fact that so much about the way the world works could be explained by a simple set of equations using calculus is undoubtedly part of what Leibniz was thinking about when he described our world as, among all possible ones, "the simplest in hypotheses and the richest in phenomena."

The replacement of classical mechanics by the newer theories of relativity and quantum mechanics only made more impressive the congruence between mathematics and physical reality. While quantum mechanics involves more sophisticated mathematics than the mechanics of Newton, it explains a much wider range of phenomena, down to the atomic level, and still does so basically using a single simple equation (Schrödinger's equation). The theory of general relativity manages to describe accurately the effects of the gravitational force on distance scales ranging from cosmological ones down to the smallest ones for which we can measure the effects of the force. It does this using the sophisticated mathematics of modern geometry, and in that language the theory can be summarized in a very simple equation. Modern physics has replaced Newtonian physics with a whole new set of fundamental concepts, but expressed in modern mathematical language, these involve very simple hypotheses and explain an incredibly rich array of phenomena.

The physicist Eugene Wigner famously wrote an article entitled "The Unreasonable Effectiveness of Mathematics in the Natural Sciences," which he ended with the following remark: "The miracle of the appropriateness of the language of mathematics for the formulation of the laws of physics is a wonderful gift which we neither understand nor deserve."[3] The standard model is today the best embodiment of the idea that given the right mathematical language, the fundamental principles of physics can be written down in extremely simple terms. This language now uses the very twentieth-century mathematics of gauge symmetry, spinors, and the Dirac equation. With this modern language one is able to write down in a few lines equations that govern all known interactions of all known particles (with an important caveat about how one treats the gravitational interaction).

Eric Baum was a fellow student of mine at Harvard and Princeton, who received his PhD in the field of quantum gravity. He later went on to work on neural networks and cognitive science, recently publishing a book on the subject entitled *What Is Thought?*[4] The argument of his book can be summarized roughly as "mind is a complex but still compact program that captures and exploits the underlying compact structure of the world." Physics tells us that the laws governing how the physical world works have an underlying simple and compact structure. Baum's argument is that our minds are fundamentally of the same nature as computer programs, programs that capture the same structure as that of the physical world and allow us to interact effectively with it. Since the sorts of distance scales with which we normally interact are those for which classical mechanics is quite accurate, it is that structure that we all share and find to be a fundamental aspect of our thought processes. The more sophisticated modern physics that comes into play at very long or very short distances is not built into our minds at a basic level, but is something that we can, with great difficulty, learn to manipulate using our more abstract mental faculties. While this is not easy, our ability to do it at all is based on the fact that there is an underlying simple structure at work, albeit one that requires the use of a highly sophisticated mathematical framework to express.

This fact that the most powerful physical theories have compact expressions in terms of the language of mathematics is what physicists generally have in mind when they refer to the beauty or elegance of these theories. Dirac famously expressed this as "If one is working from the point of view of getting beauty in one's equation, and if one has really sound insights, one is on a sure line of progress," and even more extremely, "It is more important to have beauty in one's equations than to have them fit experiment."[5] Many physicists would disagree with the second statement, but the first is one that reflects well the aspirations of many theorists.

During periods in which experiments are providing unexpected new results, the primary task of theorists is to come up with some sort of explanatory model of what the experiments are revealing, one that agrees with those already performed and that predicts what results new ones will produce. Considerations of beauty and elegance are then secondary, functioning through the principle of Occam's razor: given the many possible models that might agree with experiment, one should focus on the simplest ones. In a period such as the current one, when there are few or no unexplained experimental results, the principle that one should look for simple, beautiful theoretical explanations takes on greatly increased importance. One argument often heard for superstring theory is that it is a theory of such beauty and elegance that there must be something right about it.

This argument has always deeply puzzled me. Despite having spent a lot of time learning about superstring theory, I have never been impressed by it as something I would describe as beautiful. The quantum field theory of the standard model contains physical and mathematical ideas that are strikingly beautiful, the likes of which superstring theory cannot approach. The fundamental constructions in the standard model (the Dirac equation and the notion of a Yang–Mills field) correspond precisely to mathematical structures that are at the core of the modern viewpoint about geometry developed during the twentieth century. The conceptual structure of superstring theory asks us to believe that these are just approximate, low-energy limits of something more fundamental, but cannot tell us what this more fundamental thing is supposed to be.

Trying to find out what it is about superstring theory that some consider so aesthetically appealing, one finds various explanations by different physicists. The most superficial explanation one finds is just that the idea of getting both gravity and the standard model out of a simple easily visualizable idea is what is beautiful. The image of a vibrating string whose vibrational modes explain all known particles and forces is what many find beautiful about the theory. This sort of beauty is very much skin deep. Once one starts learning the details of ten-dimensional superstring theory, anomaly cancellation, Calabi–Yau spaces, etc., one realizes that a vibrating string and its musical notes have only a poetic relationship to the real thing at issue.

In an interview, John Schwarz explained what he meant about the beauty of superstring theory as follows:

> Q: You mentioned the beauty of the string theory. Could you elaborate a bit?
>
> A: Different people see it in different ways. A standard way of explaining this is that you have a rather simple-looking equation that explains a wide range of phenomena. That is very satisfying. Famous examples of that are due to Newton, Einstein, and so forth. We do not actually have a concise formula in string theory that explains a lot of things, so the beauty is of a somewhat different character here. What, I think, captures people in the subject is when they discover that they are dealing with a very tight mathematical system that incorporates things that nobody has understood. When you do some complicated calculation, and then you discover that the answer is surprisingly simple, much simpler than you would have expected . . .
>
> And it sort of appears as if there were some sort of mathematical miracle taking place. Of course, there are no miracles, and when you find something that looks like a miracle, that just means that there is some important concept that you haven't understood yet. When you experience that a few times, you really get seduced by the subject. That's been my experience, and I think of the other people who are now just as enthusiastic, and, maybe, even more so, than I am, have had similar experiences.[6]

Schwarz tells us that superstring theory is not beautiful in the sense that the Dirac equation is beautiful, since it is not based on a simple compelling idea that leads to powerful consequences. Instead, the theory is a complex set of interrelated constructions that when manipulated give evidence of the existence of underlying structures one still hasn't understood. Its beauty is the beauty of mystery and magic, two of Witten's suggested meanings for the M in M-theory. This kind of beauty of course may disappear without a trace once one finds out the magician's trick behind the magic or the story behind the mystery.

The initial hopes for superstring theory that motivated many to take up the subject in 1984 revolved around the anomaly cancellation conditions found by Green and Schwarz that picked out certain specific choices for the gauge group of the theory. This calculation, together with the similar but older anomaly cancellation condition that requires superstrings to live in ten-dimensional space-time, is one of the main things people have in mind when they talk about the beauty of superstring theory. It certainly would count as a beautiful mathematical explanation of the structure of the world if these cancellation conditions predicted what one sees, but space-time appears to have four, not ten, dimensions, and the gauge group of the standard model is much smaller than the gauge groups predicted by anomaly cancellation ($SO(32)$ and $E_8 \times E_8$). The hope in 1984 was that these mismatches could be accounted for by the compactification of six of the dimensions into a Calabi–Yau space. At the time, very few such Calabi–Yau manifolds were known, so it was not unreasonable to hope that one of them would do the trick and give one the standard model structure.

More than twenty years of research has shown that this was wishful thinking. There is a huge and possibly infinite number of classes of Calabi–Yau spaces, and the introduction of branes into the subject opened up vast numbers of additional new possibilities. Some superstring theorists now argue that the theory is inherently not an elegant one, but its virtue is that it can describe all sorts of very complex things, some of which are complicated enough to produce intelligent life. In his recent book *The Cosmic Landscape*, Leonard Susskind describes what happened after initial hopes for the uniqueness of string theory began to collapse:

> New possibilities kept turning up, new mathematically consistent versions of what was supposed to be a unique theory. During the 1990s the number of possibilities grew exponentially. String theorists watched with horror as a stupendous Landscape opened up with so many valleys that almost anything can be found somewhere in it.
>
> The theory also exhibited a nasty tendency to produce Rube Goldberg machines. In searching the Landscape for the Standard Model, the constructions became unpleasantly complicated. More and more

"moving parts" had to be introduced to account for all the require-
ments . . .

Judged by the ordinary criteria of uniqueness and elegance, String
Theory had gone from being Beauty to being the Beast.[7]

Susskind goes on to say that this complexity and ugliness is actually a good thing, making a peculiar argument that we will examine
in a later chapter. While he believes it is settled that superstring theory, whatever it is, has a huge and extremely complicated possible set
of vacuum states, he acknowledges that the unknown underlying
theory may be less complex or even elegant, saying, "I think I might
find the universal principles of String Theory most elegant—if I only
knew what they were."[8] He makes light of this situation as follows:

Elegance requires that the number of defining equations be small.
Five is better than ten, and one is better than five. On this score, one
might facetiously say that String Theory is the ultimate epitome of
elegance. With all the years that String Theory has been studied, no
one has ever found even a single defining equation! The number at
present count is zero. We know neither what the fundamental equations of the theory are or even if it has any. Well then, what is the theory, if not a collection of defining equations? We really don't know.[9]

Although different physicists may have varying opinions about the
beauty of superstring theory, there is little disagreement about the difficulty of the theory. The simplest version of string theory, in which
one just tries to come up with a quantization of the classical theory of
a vibrating string, already requires dealing with a host of thorny technical issues. The theory is supposed not to depend on how one para-
meterizes the string, and achieving this is rather tricky, leading to the
consistency requirement that the string must live in a 26-dimensional
space-time. This simple version of string theory doesn't appear to have
a stable vacuum, so one thing one needs to do is to consider a supersymmetric version involving fermions: the superstring. The basic
equations for the superstring are complicated and come in several different versions, with consistency now requiring 10 instead of 26
dimensions. Compactifying six of the dimensions brings in the full

complexity of the geometry of curved six-dimensional Calabi–Yau spaces, a very difficult and challenging part of modern algebraic geometry.

While ten-dimensional superstring theory compactified on a Calabi–Yau space is an extremely complex and difficult subject to master, in modern superstring theory it is just the beginning. The hope is that there is an underlying nonperturbative M-theory, but only a bewildering collection of partial results about this theory exist. Its low-energy limit is supposed to be eleven-dimensional supergravity, a rather complicated subject in itself, especially when one considers the compactification problem, which now requires understanding curved seven-dimensional spaces. Somehow, M-theory is supposed to describe not just strings, but higher-dimensional objects called branes. Many different sorts of calculations involving branes have been performed, but a fundamental theory capable of consistently describing them in an eleven-dimensional space with seven dimensions compactified still does not exist, despite more than ten years of effort looking for such a thing.

This huge degree of complexity at the heart of current research into superstring theory means that there are many problems for researchers to investigate, but it also means that a huge investment in time and effort is required to master the subject well enough to begin such research. To really understand superstring theory, one should first study quantum field theory, and in itself this is a very demanding task. Typically, graduate students take a course in quantum field theory during their second year of graduate school, in which case they can't even begin to work on superstring theory until their third year. This leaves little time to master the subject and get some new results about it in the standard four- to five-year length of a PhD program. Young superstring theorists often gain their degrees having only become really familiar with a small part of the subject, with true expertise requiring many, many years to achieve. An article about superstring theory in *Science* magazine quotes one young string theorist as follows:

> Brent Nelson, a postdoc at the University of Pennsylvania, says he read about string theory as a teenager and couldn't believe so many

people accepted something so outlandish. "I haven't learned enough," he says, "I still don't know why I should believe."[10]

Since the whole subject is so complicated and difficult, theorists trying to evaluate what is going on often rely to an unusual extent not on their own understanding of the subject, but on what others say about it. The fact that Witten took up superstring theory with such enthusiasm in 1984 had a lot to do with it becoming so popular, and his continuing belief that it remains the most promising idea to work on has a huge influence. A major reason for this is that many people rely on his judgment because they find the current state of string theory so difficult to comprehend that they are not able to reasonably form their own judgments of the situation.

Besides raising a huge barrier of entry to the subject, the difficulty of superstring theory also makes it hard for researchers to leave. By the time they achieve some real expertise they typically have invested a huge part of their careers in studying superstrings, an investment that is psychologically and professionally very difficult to give up. Philip Anderson notes,

> The sad thing is that, as several young would-be theorists have explained to me, it is so highly developed that it is a full-time job just to keep up with it. That means that other avenues are not being explored by the bright, imaginative young people, and that alternative career paths are blocked.[11]

It is also true that there are no alternatives to superstring theory that one can easily learn and quickly start researching. Other ideas remain very little developed, and many of them require dealing with a whole slew of different mathematical ideas that are not part of a physicist's normal training. Mathematicians don't make things any easier, since readable expository material about much of modern mathematics is sorely lacking. The culture of mathematics values highly precision, rigor, and abstraction, not the sort of imprecise motivational material and carefully worked out examples that make a subject accessible to someone from the outside trying to get some idea of what is going on. This makes the research literature often impenetrable to all but those already expert in a field. There is often a somewhat intellectually macho attitude among some mathematicians, an attitude that since they overcame great hurdles to understand something, there's no reason to make it easier and encourage others less talented and dedicated than themselves.

However, this sort of arrogance among mathematicians pales in comparison with the degree of arrogance one sometimes encounters among superstring theorists. They often seem to be of the opinion that only real geniuses are able to work on the theory, and that anyone who criticizes such work is most likely just too stupid and ignorant to understand it. There is a striking analogy between the way superstring theory research is pursued in physics departments and the way postmodern "theory" has been pursued in humanities departments. In both cases, there are practitioners that revel in the difficulty and obscurity of their research, often being overly impressed with themselves because of this. The barriers to understanding that this kind of work entails make it very hard for any outsiders to evaluate what, if anything, has been achieved.

The level of complexity and difficulty of superstring theory is probably simply an indication that the subject is on the wrong track, and a reflection of the fact that no one has any idea whether there really is some unknown simple fundamental M-theory. In successful physical theories such as the standard model, the ideas involved can be difficult for a student to absorb, but once one gets to a certain point the foundations are clear and one can see how the structure and implications of the theory follow from certain basic assumptions. The unsatisfying nature of some aspects of the standard model leads us to believe that there is some more fundamental structure behind it that we still don't understand. Presumably, once someone figures out what this is, it will not be especially more difficult for others to learn than the standard model itself. Finding such a new, deeper, and better way of thinking about fundamental physics is, however, an intellectually extremely demanding task. Unfortunately, it is not at all inconceivable that it is one that is beyond the capabilities of human beings if they are unaided by clues from experimentalists.

Is Superstring Theory Science?

40 major qualities of the Unified Field that can be equated with the characteristic qualities of the 40 aspects of the Veda and Vedic Literature have been derived from the mathematical formula of the Unified Field as given by Superstring Theory.

—MAHARISHI MAHESH YOGI,
UNIFIED FIELD OF ALL THE LAWS OF NATURE

No matter how things turn out, the story of superstring theory is an episode with no real parallel in the history of modern physical science. More than twenty years of intensive research by thousands of the best scientists in the world producing tens of thousands of scientific papers has not led to a single testable experimental prediction of the theory. This unprecedented situation leads one to ask whether one can really describe superstring theory research as scientific research in the field of physics. This question tends to take on two different forms. One form of the question that many physicists ask is whether superstring theory should not perhaps be described as mathematics rather than physics. A second form of the question asks whether the theory is a science at all.

Since I spend most of my time in a mathematics department, it's very clear to me how my mathematician colleagues would answer the question whether superstring theory is mathematics. They would uniformly say, "Certainly not!" Mathematicians see the defining activity of their discipline to be the making of precise statements of theorems about abstract mathematical entities and the construction of rigorous proofs of these theorems. The fact that superstring theory research refers to speculative physical entities is not really a problem, since mathematicians are masters of abstraction, and can easily change any precise theoretical framework into one expressed in the language of abstract mathematics. The problem is that superstring theory is not really a theory, but rather a set of hopes that a theory exists. To a mathematician, a set of hopes that a theory exists, hopes that come purely out of physical motivations, is definitely not mathematics at all. Just as in physics, such a set of hopes can in principle be used as motivation for making a precise set of conjectures about what is true, but until the conceptual framework reaches the point of one's being able to do this, it is not clear how one can really use it.

On the other hand, many physicists who don't work on superstring theory often characterize it as being mathematics. In the majority of cases, this is meant as a negative characterization, since many physicists share the attitude Gell-Mann once held that abstract mathematics is some form of self-abuse. Superstring theory is to a large degree thought of by mainstream physicists as mathematics and by mainstream mathematicians as physics, with each group convinced that it makes no sense within their frame of reference but presumably does within someone else's.

A favorite decoration for particle theorists' office doors during the mid-1980s was a very large and very colorful poster distributed by the Maharishi International University. The poster included the basic equations of eleven-dimensional supergravity, annotated with a detailed explanation of the relation of each term in the equations to the Maharishi's version of Indian philosophy. The epigraph at the beginning of this chapter is from a newer, equally colorful document, now distributed online by the Maharishi University of Management. The new document is more up to date (supergravity has been replaced by superstring theory), but otherwise it seems to be much the same sort of thing.

The main person behind all this is not the Maharishi, but a physicist named John Hagelin. Hagelin was a graduate student in particle theory at Harvard during the late 1970s, and I remember attending a quantum field theory class with him at that time. Since a roommate of mine had a certain interest in Transcendental Meditation (TM) and knew Hagelin from the local TM center, somehow I got around to talking to him. His interest in quantum field theory seemed to have a bizarre side to it, since he wanted to use it to explain how TM adepts were able to levitate, but in other ways he behaved much like other graduate students.

Hagelin graduated from Harvard a couple of years later with several serious papers on particle theory to his name, and then went to SLAC as a postdoc for a few years. During this time he was working on supersymmetric extensions of the standard model and grand unified theories, collaborating with many of the leading figures in that field and writing a large number of papers, some of which are frequently cited to this day. By 1984, Hagelin had left SLAC and moved to the Maharishi International University, in Fairfield, Iowa, and had begun to build up a physics department there. At that time, particle theory departments would print up large numbers of copies of their members' new articles in preprint form and mail them around to other departments. I recall seeing several Maharishi International University preprints on particle theory from that period. In content they were indistinguishable from many other preprints on similar topics, but they were somewhat unusual in that they were printed on pink rather than white paper.

During the rest of the 1980s and early 1990s, Hagelin continued to produce mainstream scientific articles, now often trying to work out the implications of various grand unified theories derived from string theory. By 1995, Hagelin had written 73 scientific articles, most of them published in very prestigious particle theory journals, many of them cited by more than a hundred later articles. If one examines the list of these articles in the SLAC database, a couple of titles stand out: "Is Consciousness the Unified Field? (A Field Theorist's Perspective)" and "Restructuring Physics from Its Foundations in Light of Maharishi's Vedic Science." Looking at these articles, one finds that from the mid-1980s on, Hagelin was identifying the "unified field" of superstring theory with the Maharishi's "unified field of consciousness." Maharishi International University was requiring all its first-year students to take a "twenty-lesson introduction to the conceptual foundations of unified field theories," in which presumably the connection between superstring theory and consciousness was explained in detail. In recent years, Hagelin has stopped writing physics papers and has achieved notoriety as the presidential candidate of the "Natural

Law Party," most recently promoting the idea of fighting terrorism with a "new Invincible Defense Technology based on the discovery of the unified field."

Virtually every theoretical physicist in the world rejects all of this as utter nonsense and the work of a crackpot, but Hagelin's case shows that crackpots can have PhDs from the Harvard Physics Department and a large number of frequently cited papers published in the best peer-reviewed journals in theoretical physics. How does the field protect itself against crackpots? Even though Hagelin undoubtedly sees his work as all of a piece, how can one separate out what is legitimately science from what is irrational wishful thinking on his part?

Human beings engage in many different sorts of attempts to explain the world around them, but only a specific sort of explanation is normally considered to be scientific. An explanation that allows one to predict successfully in detail what will happen when one goes out and performs a feasible experiment that has never been done before is the sort of explanation that most clearly can be labeled "scientific." Explanations that are grounded in traditional or religious systems of belief and that cannot be used to predict what will happen are the sort of thing that clearly does not deserve this label. This is also true of explanations based on wishful thinking or ideology, where the source of belief in the explanation can be identified as something other than rational thought.

The question whether it is possible to decide what is science and what is not, and if so how to make this decision, is a central issue in the field of philosophy of science. The best-known proposed criterion for deciding what is science and what isn't is the criterion of falsifiability generally attributed to the philosopher Karl Popper. By this criterion an explanation is scientific if it can be used to make predictions of a kind that can be falsified, i.e., can be shown to be wrong. The falsifiability criterion can in some circumstances be slippery, because it may not always be clear what counts as a falsification. Observations may be theory-laden, since some sort of theory is needed even to describe what an experiment is seeing, but this problem doesn't seem to be at issue in this context.

While specific models can be straightforwardly falsifiable, the question whether one can falsify a more general theoretical framework is more subtle. Over the years, many preliminary experimental results disagreeing with standard-model predictions have been reported. In each of these cases it was generally possible to come up with an extension of the standard model that agreed with the new results, but at the cost of significantly increasing the complexity of the theory. None of these experimental results ever held up, with more careful analysis always showing that no extension of the standard model was actually needed. Given some theoretical framework, one can often find a way of fitting all sorts of different experimental results if one allows oneself to use arbitrarily complicated models within that framework. Aesthetics comes into the problem of whether a given framework is falsifiable, since one has to restrict oneself to considering relatively simple and natural models within the framework. If one allows extremely complex and baroque constructions, one can often get agreement with just about any experimental result.

The standard model is an excellent example of a falsifiable theory, since it is one of the simplest possible models of its kind, and it can be used to generate an infinite set of predictions about the results of particle physics experiments, all of which in principle can be checked in an unambiguous way. On the other hand, superstring theory is at the moment unarguably an example of a theory that can't be falsified, since it makes no predictions. No one has come up with a model within the superstring theory framework that agrees with the known facts about particle physics. All attempts to do so have led to very complicated constructions that show every sign of being the sort of thing one gets when one tries to make an inappropriate theoretical framework fit experimental results. At the same time, due to the lack of a nonperturbative theory, the superstring theory framework remains too poorly understood for anyone to be completely sure about what sorts of truly consistent models fit into it.

Speaking at a conference in 1998 attended mostly by experimental physicists, the superstring theorist Joseph Polchinski stated, "I am sure that all the experimentalists would like to know, 'How do I falsify string theory? How do I make it go away and not come back?' Well you can't. Not yet."[1]

By the falsification criterion, superstring theory would seem not to be science, but the situation is more complex than that. The tricky issue is Polchinski's "not yet." Much theoretical activity by scientists is speculative, in the sense that it consists in asking questions of the kind, "What if I assume X is true. Could I then construct a real theory using this assumption?" This is certainly the kind of thing scientists spend a lot of time doing, and one presumably doesn't want to label it unscientific. Superstring theory is very much a speculative endeavor of this kind. Theorists involved in this area are considering a very speculative assumption: that one should replace the notion of elementary particle with strings or more exotic objects, and trying to see whether a scientific theory capable of making falsifiable predictions can be built on this assumption.

Generalizing the notion of "scientific" to include speculation of this kind would definitely make superstring theory a science. But does one really want to say that all such speculative activity is scientific? A favorite story among cosmologists goes as follows (this version is due to Stephen Hawking, but there are many others):

A well-known scientist (some say it was Bertrand Russell) once gave a public lecture on astronomy. He described how the Earth orbits around the Sun, and how the Sun, in turn, orbits around the center of a vast collection of stars called our galaxy. At the end of the lecture a little old lady at the back of the room got up and said: "What you have told us is rubbish. The world is really a flat plate supported on the back of a giant tortoise." The scientist gave a superior smile before replying, "What is the turtle standing on?" "You're very clever, young man, very clever," said the little old lady. "But it's turtles all the way down."[2]

While physicists enjoy this story and its many variants (William James or Einstein often replaces Russell), another version of it is well known to anthropologists. Theirs is due to Clifford Geertz and goes as follows:

There is an Indian story—at least I heard it as an Indian story—about an Englishman who, having been told that the world rested on a platform which rested on the back of an elephant which rested in turn on the back of a turtle, asked (perhaps he was an ethnographer; it is the way they behave), what did the turtle rest on? Another turtle. And that turtle? "Ah, Sahib, after that it is turtles all the way down."[3]

Geertz tells the story to make a point about "antifoundationalism" and goes on to write,

Nor have I ever gotten anywhere near to the bottom of anything I have ever written about, either in the essays below or elsewhere. Cultural analysis is intrinsically incomplete. And, worse than that, the more deeply it goes the less complete it is.

Particle theory, unlike ethnography, is very much a science that is supposed to have a foundation, and the more deeply one gets into this foundation, the more complete the theory is supposed to be. The standard model is a theory that provides a foundation for the prediction and understanding of a wide range of phenomena. Current research is supposed to be focused on both shoring up those places where the foundation is a bit shaky, and finding an even more complete theory.

To choose a very slightly less unreasonable form of the turtle theory, what if I decide to speculate that at sufficiently short distance scales, physics is to be described not in terms of particles, strings, etc., but in terms of turtles? If I were to announce that I was investigating the prospects of a unified theory built on the assumption that the world is constructed out of extremely small turtles, and that from this assumption I hoped to derive the standard model and calculate its undetermined parameters, most people would say that I was not doing science. On the other hand, if after a few months work I did manage to derive the parameters of the standard model from the turtle assumption, then people would have to change their minds and admit that, yes, I had been doing science all along. So the question whether a given speculative activity is science seems not to be one admitting an absolute answer, but instead is dependent on the overall belief system of the scientific community and its evolution as scientists make new theoretical and experimental discoveries. Speculative research on a problem using an approach characterized as unreasonable and unworkable by most scientists who have thought long and hard about the problem probably should not be called scientific research, especially if it goes on and on for years showing no sign of getting anywhere. On the other hand, if a large part of the scientific community thinks a speculative idea is not unreasonable, then those pursuing this speculation must be said to be doing science.

The speculation known as superstring theory continues to qualify as a science by this criterion, since a large fraction of theorists consider it a reasonable assumption worth trying out. The decision to call it such is very much a social one, and it has its basis in the shared judgment of many, but not all, physicists. In the case of superstring theory, there are many physicists who believe that the speculative assumptions involved are probably wrong and even more likely to be such that one simply can never hope to turn them into a theory that can be used to make predictions. Superstring theorists are well aware that this is an issue of contention in the physics community, and that if the theory continues to be unable to predict anything, at some point the community will stop being willing to call what they do science.

The qualms that many scientists have about superstring theory are often expressed as the worry that the theory may be in danger of becoming a religion rather than a science. Glashow is one physicist who has expressed such views publicly:

> Perhaps I have overstated the case made by string theorists in defence of their new version of medieval theology where angels are replaced by Calabi–Yau manifolds. The threat, however, is clear. For the first time ever, it is possible to see how our noble search could come to an end, and how Faith could replace Science once more.[4]

I have heard another version of this worry expressed by several physicists, that superstring theory is becoming a cult, with Witten as its guru. For an example of this, recall Magueijo's comments on M-theory quoted earlier. Some string theorists do express their belief in string theory in religious terms. For instance, a string theorist on the faculty at Harvard used to end all his e-mail with the line, "Superstring/M-theory is the language in which God wrote the world." String theorist and author Michio Kaku, when interviewed on a radio show, described the basic insight of string theory as, "The mind of God is music resonating through 11-dimensional hyperspace."[5] Some physicists have joked that, at least in the United States, string theory may be able to survive by applying to the federal government for funding as a "faith-based initiative." In recent years, the Templeton Foundation, a foundation dedicated to promoting a rapprochement between science and religion, has been supporting conferences that have featured many prominent string theorists. Glashow's worry about the possibility of theology replacing science sometimes seems a very serious one.

Personally, I don't think the categories of cult or religion are especially appropriate in this circumstance, since they refer to human activities with many quite different characteristics from those that typify what is going on in the physics community. On the other hand, as years go by and it becomes clear that superstring theory has failed as a viable idea about unification, the refusal to acknowledge this begins to take on ever more disturbing implications. We have seen that there is no way to separate clearly the question of what is and isn't science from the very human issues of what people choose to believe and why. Science thus has no grant of immunity from some of the dangers of cultish behavior to which human beings can fall prey. Strong internal norms of rationality are needed and must be continually enforced to ensure that a science continues to deserve that name.

The Bogdanov Affair

While I was in the process of writing this book, one morning in October 2002 I came into the office and began the day as usual by reading my e-mail. A couple of physicist friends had forwarded to me reports of a rumor, one that they knew I would find interesting. The rumor was that two French brothers, Igor and Grichka Bogdanov, had concocted what some people were calling a "reverse-Sokal" hoax. In 1996, the physicist Alan Sokal had written a carefully constructed but utterly meaningless article with the title "Transgressing the Boundaries: Toward a Transformative Hermeneutics of Quantum Gravity." The article contained no rational argument and instead strung together unsupported claims, breathtaking leaps of logic, and a large collection of the sillier parts of the writings of both postmodern theorists and some scientists. It ended up making no sense at all, but was side-splittingly funny (if you were in on the joke). Sokal submitted the article to the well-known and rather prestigious academic journal *Social Text,* whose editors accepted it for publication in an issue on "Science Studies." The rumor in my e-mail was that the Bogdanov brothers had done something similar, constructing as a hoax utterly meaningless articles about quantum gravity, then getting them accepted by several journals and even using them to get a French university to award them PhDs.

After Sokal's hoax first appeared, I had thought fairly seriously about the idea of trying to write a superstring theory paper as a hoax, and seeing whether I could get it published in a physics journal. If I started with one of the more complicated and incoherent articles on superstring theory, reworked the argument to add a new layer of incoherence and implausibility and a few clever jokes, the result would be something that made no sense at all, but perhaps could pass many journal editors and referees. After thinking about this for a while, I finally gave up on the project because it was unclear to me what I could claim to have proved if successful. Sokal's opponents had pointed out at the time that he had constructed what they would describe as a not very good argument of a kind that they endorsed, and whether he himself believed it was irrelevant. Similarly, any superstring theory hoax on my part could be characterized as a not very good piece of superstring theory research that I had managed to get by overworked and inattentive referees. The fact that I did not believe what I had written would prove nothing.

That morning I looked up the Bogdanov brothers' theses on the web and quickly skimmed through them. They didn't look like a hoax. In particular, very much unlike Sokal's paper, there was nothing at all funny about them. Later that afternoon I heard fresh rumors that a *New York Times* reporter had contacted one of the brothers, who had indignantly denied any hoax. It seemed that this was just one more example of incompetent work on quantum gravity, something not especially unusual.

The next day, many e-mails were being forwarded around about the Bogdanov "hoax." For example, someone who was visiting the Harvard string theory group sent a friend of his the following report:

> So no one in the string group at Harvard can tell if these papers are real or fraudulent. This morning, told that they were frauds, everyone was laughing at how obvious it is. This afternoon, told they are real professors and that this is not a fraud, everyone here says, well, maybe it is real stuff.

This ultimately reached one of the Bogdanov brothers, who circulated it widely in an e-mail denying the existence of a hoax. Since I had some free time, I decided to look more closely at the two theses. One of them, Grichka's, was a pretty impenetrable piece of work mostly in the area of quantum algebra, something about which I am not particularly knowledgeable. The other, Igor's, was mostly about topological quantum field theory, a field I know much better. Igor's thesis was rather short, and a large part of it was an appendix consisting of four of his published papers. Looking carefully at these papers, I immediately noticed that two of them were nearly identical, including word for word identical abstracts, and both seemed to be extracts from one of the others. Upon further investigation, it turned out there was a fifth paper the brothers had published in a different journal that was again more or less identical to the two others.

This certainly caught my attention, since while lots of people write incoherent papers, I had never heard of anyone ever engaging in this kind of extreme self-plagiarism by getting nearly identical papers published in three different journals. Looking more carefully at the longest of their papers, the one from which three others had been extracted, it became clear that it was a rather spectacular piece of nonsense, a great deal more so than anything I had previously seen in a physics journal. The introduction was an impressive array of invocations of various ideas, many of them about topological quantum field theory, but pretty much all of them either meaningless or simply wrong. The body of the article was no better, containing many completely ludicrous statements. The whole thing was funny, but it was looking more and more as if this was unintentional.

Considered as a whole, what the Bogdanov brothers had managed to do (besides getting their theses accepted) was to publish five articles, three of which were nearly identical, in peer-reviewed journals. Two of the journals were quite well known and respected (*Classical and Quantum Gravity* and *Annals of Physics*), while a third was one with an illustrious history, but where standards were known to have slipped in recent years (*Nuovo Cimento*), and the final two were more obscure (*Czechoslovak Journal of Physics* and *Chinese Journal of Physics*). Evidently, five sets of editors and referees had gone over these papers and accepted them for publication, without noticing that they were egregious nonsense. Later on, several of the referees' reports surfaced, two of which were quite perfunctory, but one of which was much more detailed, making seven recommendations about changes that needed to be made to the paper before it would be suitable for publication. Ultimately, one of the journals involved (*Classical and Quantum Gravity*) released a statement saying that its editorial board had agreed that publication of the paper was a mistake and (undisclosed) steps would be taken to keep this from happening again. The editor of one of the other journals (*Annals of Physics*), Frank Wilczek, also said that publication had been a mistake, one made before he had become editor, and he hoped to improve the standards of the journal.

Various journalists looked into the story, and articles about the Bogdanovs were published in several places, including the *Chronicle of Higher Education*, *Nature*, and the *New York Times*. Many details emerged about the brothers and how they got their PhDs. They are in their fifties, had a TV show in France during the 1980s involving science fiction, and now have a new show of short segments in which they answer questions about science. Moshe Flato, a mathematical physicist at the Université de Bourgogne in Dijon, had agreed to take them on as students in the early 1990s, but had died unexpectedly in 1998. After his death the brothers presented their theses, and one of them (Grichka) was passed and awarded a mathematics PhD in 1999. The second (Igor) was failed, but told he could try again if he could get three articles accepted by peer-reviewed journals, something he went ahead and did, as we have seen. He was finally also passed and given a physics PhD in 2002.

It is hard to give anything like a summary of the Bogdanov papers, since they make so little sense, but roughly they claim to be saying something about the beginning of space and time using topological quantum field theory, and all this is somehow related to quantum field theory at high temperature. The discussion section at the end of their three identical papers is all about relations of their work to superstring theory and the problem of supersymmetry breaking. To get an idea of what the referees thought of the papers, here is the only substantive paragraph in one of the referees' reports:

> Motivated by string theory results, in this paper the author discussed the space-time below Planck scale as a thermodynamic system subject to KMS condition. Since the physics of the Planck scale has been largely unexplored, the viewpoint presented in this paper can be interesting as a possible approach of the Planck scale physics.

The significance of the Bogdanov affair was hotly debated among physicists for the next few months, with most superstring theorists taking the position that this was just a case of a few referees being lazy, and that these weren't papers about superstring theory anyway.

While very few in the particle theory community have tried to defend the Bogdanovs' work or to claim that it makes much sense, some very weird e-mails did make the rounds. One superstring theorist circulated to his colleagues an attack on a mathematical physicist who had pointed out evidence that the Bogdanovs did not understand what a topological quantum field theory is, making clear in the process that he himself shared the brothers' misconception.

The Bogdanovs wrote to me politely in February 2003, defending their work and asking me what I thought was wrong with it. I made the mistake of thinking that they could perhaps use some helpful advice and wrote back a friendly response. In it I mainly tried to make the point that what they had written was too vague and incoherent to make much sense, and that they needed to make their ideas much clearer and more precise before anyone could tell whether they had any value.

Late in 2003 I received an e-mail from a Professor Liu Yang, supposedly at the International Institute of Theoretical Physics in Hong Kong, defending in detail the work of the Bogdanovs in the field of Riemannian Cosmology. Upon investigation, it became clear that there is no such institute; nor is there such a Professor Yang. Looking closely at the e-mail header showed that it had come from a computer attached to a dial-up connection in Paris, but configured to claim a Hong Kong Internet address. I did not pay much attention to this, but it convinced me the Bogdanovs were not the innocent, guileless sorts that I had previously thought.

Early in June 2004 the Bogdanovs published a book in France with the title *Avant le Big-bang* (Before the Big Bang), which sold quite well. In their book they used part of the e-mail I had sent them the year before to claim that I was now a supporter of theirs. They mistranslated one line of my e-mail (where I was being too polite), "It's certainly possible that you have some new worthwhile results on quantum groups," as "Il est tout à fait certain que vous avez obtenu des résultats nouveaux et utiles dans les groupes quantiques" (It is completely certain that you have obtained new worthwhile results on quantum groups).

Around this same time a message defending the Bogdanovs appeared from a "Roland Schwartz," whose computer was using exactly the same Paris Internet service provider as Professor Yang. Later that month, the brothers started sending e-mails using an Internet domain name purporting to be an International Institute of Mathematical Physics in Riga. This address hosts a website for a Mathematical Center of Riemannian Cosmology, devoted to the work of the Bogdanovs. In a posting on a French Internet newsgroup, the brothers helpfully explain that the University of Riga set up the site for them, and that's why it has a Lithuanian domain name. One problem with this is that Riga is in Latvia, not Lithuania. I take this rather personally, since my father was born in Riga (the Latvian version of my name is Voits). He and his parents became exiles at the time of the Soviet occupation during World War II. I have visited Riga several times (including a visit to the university); the first time was soon after independence, on a trip with my father while he was still alive. Riga is a beautiful city, with the downtown not much changed since before the war. In recent years, the old city and much of the downtown have been elegantly renovated, and Riga is now once again a large, vibrant city with great restaurants, hotels, and shops. I am sure, however, that it does not have an International Institute of Mathematical Physics.

Leaving aside the issue of whether the Bogdanovs are hoaxers or really believe in their own work, this episode definitively showed that in the field of quantum gravity one can easily publish complete gibberish in many journals, some of them rather prominent. Whereas Sokal put a lot of effort into fooling the *Social Text* editors, the nonsensical papers of the Bogdanovs may have been guilelessly produced, and then made it into five journals, not just one. This brings into question the entire recent peer-reviewed literature in this part of physics, since the refereeing process is apparently badly broken.

One unusual thing about the Bogdanov papers is that they were never submitted to the online preprint database used by virtually all particle theorists and most mathematicians. Fewer and fewer physicists ever look at print journals these days, since essentially all recent papers of interest are available conveniently on the Web from the database. The continuing survival of the journals is somewhat mysterious, especially since many of them are very expensive. A typical large university spends over $100,000 a year buying physics journals, the content of which is almost all more easily available online for free. The one thing the journals do provide that the preprint database does not is the peer-review process. The main thing the journals are selling is the fact that what they publish has supposedly been carefully vetted by experts. The Bogdanov story suggests that, at least for papers in quantum gravity in some journals, this vetting is no longer worth much. Another reason for the survival of the journals is that they fulfill an important role in academia, where too often the main standard used to evaluate researchers' work is the number of their publications in peer-reviewed journals, something that was at work in the decision to pass Igor Bogdanov's thesis. The breakdown in refereeing is thus a serious threat to the whole academic research enterprise.

Why did the referees in this case accept for publication such obviously incoherent nonsense? One reason is undoubtedly that many physicists do not willingly admit that they don't understand things. Faced with a stew of references to physics and mathematics in which they were not expert, instead of sending it back to the editor or taking the time to look closely into what the authors were saying, the referees decided to assume that there must be something of interest there, and accepted the articles with minimal comment. The referee's report reproduced earlier shows clearly the line of thinking at work: "Well, this somehow has to do with string theory, quantum gravity, and the beginning of the universe, and it uses something called the 'KMS condition,' which is supposed to be important. Nothing published in this whole area really makes complete sense, so maybe this is no worse than lots of other stuff and maybe there's even an intelligible idea in here somewhere. Why not just accept it?"

The Bogdanov affair convincingly shows that something is seriously broken in that part of the scientific community that pursues speculative research in quantum gravity. A sizable number of referees and editors were not able to recognize complete nonsense for what it was, or if they were capable of doing so, felt that it was just not worth the trouble. The theoretical physics community seems so far to have reacted to this episode by trying to deny or minimize its significance, thus ensuring that the problems it highlights will continue for the foreseeable future.

The Only Game in Town

The Power and the Glory of String Theory

> A guy with the gambling sickness loses his shirt every
> night in a poker game. Somebody tells him that the
> game is crooked, rigged to send him to the
> poorhouse. And he says, haggardly, "I know, I know.
> But it's the only game in town."
>
> —KURT VONNEGUT, *THE ONLY GAME IN TOWN*

When talking to many superstring theorists about why they continue to work on the theory despite its continuing lack of any success in reaching its goals, the most common justification I have heard is some version of, "Look, it's the only game in town. Until someone comes up with something else more promising, this is where the action is." This kind of justification has been very much in evidence since the first superstring revolution in 1984, when so many researchers started working on the subject. In an interview in 1987, David Gross made the following comments about the reasons for the popularity of superstring theory:

> The most important [reason] is that there are no other good ideas
> around. That's what gets most people into it. When people started to
> get interested in string theory they didn't know anything about it. In
> fact, the first reaction of most people is that the theory is extremely
> ugly and unpleasant, at least that was the case a few years ago when
> the understanding of string theory was much less developed. It was
> difficult for people to learn about it and to be turned on. So I think
> the real reason why people have got attracted by it is because there is
> no other game in town. All other approaches of constructing grand

unified theories, which were more conservative to begin with, and only gradually became more and more radical, have failed, and this game hasn't failed yet.[1]

Gross was and is very much an enthusiast for superstring theory, unlike many other physicists who from the beginning found it a not very plausible idea, and had a difficult choice to make about whether to work on it. The science writer Gary Taubes at the end of his book *Nobel Dreams* tells of the following conversation with another particle theorist:

On August 4, 1985, I sat in the cantina at CERN drinking beer with Alvaro de Rujula . . . De Rujula predicted that 90 percent of the theorists would work on superstrings and the connection with supersymmetry, because it was fashionable. When he intimated that this was not a healthy state, I asked him what he would prefer to work on. Rather than answer directly, he digressed.

"It must be remembered," de Rujula told me, "that the two people most responsible for the development of superstrings, that is to say Green and Schwarz, have spent ten to fifteen years systematically working on something that was not fashionable. In fact they were ridiculed by people for their stubborn adherence to it. So when people come and attempt to convince you that one must work on the most fashionable subject, it is pertinent to remember that the great steps are always made by those who don't work on the most fashionable subject."

"The question then," I said, "is what do you work on instead? What will your next paper be on?"

"That's a question for each theorist to ask himself," he replied. "And it depends on whether you want to survive as a theorist, or you have the guts to think that pride in your own work is more important than the momentary recognition of your fashionable contribution. That's for each person to decide by himself, depending on his level of confidence in his own genius."

"So," I repeated, "what is your next paper going to be on?"

"I'm trying to tell you," de Rujula said, "that I have no idea."[2]

The fact that superstring theory research was the only game in town in the mid-1980s, at a time when there had not been a great deal of work on the theory, and one could still reasonably hope that it would lead to great things, is not very surprising. It is much harder to understand why it continues to be the only game in town more than twenty years later, in the face of ever increasing evidence that it is fundamentally a research program that has failed.

In 2001, after I posted a short article on the physics preprint archive evaluating the situation of superstring theory and strongly making the point that it was now clear that the idea had been a failure, I very quickly got a lot of e-mail in response. The only critical messages came from two superstring theory graduate students, who were of the opinion that I was an incompetent idiot threatening to hold back the progress of science. A huge number of congratulatory messages arrived, many with an aspect that surprised me. These messages remarked on my courage and expressed the hope that I would survive what they expected to be a fierce personal attack from superstring theorists. I hadn't known that so many people in the physics community not only were skeptical of superstring theory, but even felt that the subject was perpetuating itself through some sort of intimidation. My position in a mathematics department is such that I have little to worry about in terms of professional retaliation, but many of my correspondents felt very differently, one of them even referring to superstring theorists as a "mafia." This gave a different color to the "only game in town" characterization. Many physicists seemed to feel that anyone who threatened the successful operation of the superstring theory game might need to worry about his or her professional safety.

Except for the two overly enthusiastic graduate students, the response from superstring theorists to my initial article was perfectly polite. The only sort of intimidation I experienced was an intellectual one, since many of the leading figures in this field are brilliant, hardworking, very talented, and with huge accomplishments to their credit. To come to the conclusion that what they are doing is seriously wrongheaded was not at all easy. By far the most common reaction from superstring theorists has been to ignore my arguments on the grounds that I wasn't saying anything not well known to people in the field. While some of my friends and colleagues who work in this area undoubtedly found my arguments churlish, they also knew that the problems I was discussing were real. Many of them very much hope for new ideas to appear, for the day to come when superstring theory is no longer the "only game in town."

More recently, as I have found the continuing dominance of superstring theory in particle physics taking on an increasingly disturbing aspect, I have been engaged in two different projects to draw wider attention to this problem. The first is the current book, which I started writing in 2002. The second is a weblog, begun early in 2004, where I have been posting information about topics in mathematics and physics that I think others might find interesting, including a lot of critical material about the latest developments in superstring theory. These two projects have generated significantly more negative reaction than my earlier article, undoubtedly because they have been more difficult to ignore.

The weblog has generated far more interest than I ever expected. Currently, the main page gets more than 8,000 connections per day from all over the world. Many of the readers of the weblog are superstring theorists; one indication of this is the large number of connections from computers at academic institutions that have "string" as part of their names. During the "Strings 2004" and "Strings 2005" conferences in Paris and Toronto, the webserver logs showed several connections from the wireless access point in the conference lecture hall. It appears that during some of the more boring talks, more than one string theorist was checking to see what I and others had to say on the weblog.

The weblog includes a comment section, and some superstring theorists have chosen to use this to attack me personally, responding to some of my specific scientific criticisms of superstring theory by attacking me as ignorant and incompetent. One of the more excitable such superstring enthusiasts, a Harvard faculty member, even at one point used this comment section to express the opinion that those who criticized the funding of superstring theory were terrorists who deserved to be eliminated by the U.S. military. I'm afraid he seemed to be serious about this.

An earlier version of the book you are reading was originally considered for publication by Cambridge University Press, beginning early in 2003 after I met an editor there who expressed an interest in the manuscript. While Cambridge also publishes many of the best-known books on string theory, the editor seemed interested in publishing what I had written, partly to provide some balance on the topic. The manuscript was sent out to referees, and I was optimistic about the outcome, since I was confident that it contained no major errors, and so should easily pass a standard academic refereeing process. I suppose I expected non-string-theorist referees to have a positive reaction, and string theorist referees to acknowledge that I had my facts right even if they disagreed with some of my conclusions.

While the reports from some of the referees were very positive and strongly endorsed publication by Cambridge, the behavior of the string theorists was not what I had expected. In the first round of refereeing, someone described to me by the editor as "a well-known string theorist" wrote a short report claiming that the manuscript was full of errors, but would give only one example. He or she then took a sentence I had written out of context, misquoted it, turning a singular into a plural, and construed this misquoted sentence as evidence that I didn't know about certain developments in the field. As written and in context, this sentence was a perfectly accurate summary of the state of knowledge about a certain problem in quantum field theory. The report then ended, "I could write a long criticism of the manuscript, but that really shouldn't be necessary. I think that you would be very hardpressed to find anybody who would say anything positive about this manuscript." I was sent a copy of this report in the same e-mail as another report enthusiastically endorsing publication. That referee also compared my criticisms of string theory to criticisms of the teaching of evolution by creationists.

Before I saw this report I was somewhat worried about a few of the things that I had written, feeling that they came too close to accusing string theorists of intellectual dishonesty. After seeing this report, I stopped being much concerned about that. Clearly the level of such dishonesty and the extent to which many string theorists were unwilling to acknowledge the problems of their subject were far beyond anything that I had originally imagined. The Cambridge editor agreed that the first referee's report lacked credibility and in some ways provided evidence for problems my manuscript was addressing, but felt that he could not go ahead with publication without more positive reports.

A second round of refereeing produced another very positive report (from a physicist who has worked on string theory), but also another very negative one. This second one (also from a physicist who has worked on string theory) did not claim that I had any of my facts wrong, even saying that the referee shared some of my critical views about string theory. The report was rather short, but it expressed forcefully the view that the leaders of the particle theory community could take care of their own problems, that I had no business criticizing them, and strongly recommended against publication by Cambridge. The editor offered to try a third round of refereeing, but by now it was clear to me that even if they couldn't answer my arguments, string theorists would strongly oppose publication. I would be wasting my time pursuing this further with Cambridge, since they were unlikely to publish something vehemently opposed by leading figures in the field, even if this opposition was not backed up by any scientific argument. I then sent the manuscript off to editors at several other university presses, but the results were again negative, with two editors explicitly telling me that, while what I had written was very interesting, it was simply too controversial for publication by a university press.

The extent to which superstring theory is "the only game in town" is hard to exaggerate, as is the triumphalist attitude of some of its practitioners. At a meeting of the American Academy for the Advancement of Science in 2001, David Gross gave a talk entitled "The Power and the Glory of String Theory," which gives an idea of the tone of many superstring theory talks.[3] As another example, Joseph Polchinski, in his 1998 SLAC summer school lectures, began one lecture by saying, "On Lance Dixon's tentative outline for my lectures, one of the items was 'Alternatives to String Theory.' My first reaction was that this was silly; there are no alternatives."[4]

Many string theorists seem to take the attitude that it is inconceivable that superstring theory is simply a wrong idea. They feel that while the current version of superstring theory may not be right, it must somehow be a large part of whatever the ultimately successful theory will be. In an article about Witten and his superstring theorist colleagues at the Institute for Advanced Study in Princeton, one of them is quoted as follows: "'Most string theorists are very arrogant,' says Seiberg with a smile. 'If there is something [beyond string theory], we will call it string theory.'"[5]

At the moment, the director of the institute (Peter Goddard) is a string theorist, as are all of the permanent physics faculty except for Stephen Adler, a theorist appointed in the late 1960s and now nearing retirement.

In a *New York Times* article written in 2001 entitled "Even Without Evidence, String Theory Gains Influence," science reporter James Glanz wrote,

> Scientists have yet to develop more than fragments of what they presume will ultimately be a complete theory.
>
> Nevertheless, string theorists are already collecting the spoils that ordinarily go to the experimental victors, including federal grants, prestigious awards and tenured faculty positions. Less than a decade ago, there were hardly any jobs for string theorists, said Dr. David Gross . . .
>
> "Nowadays," Dr. Gross said, "if you're a hotshot young string theorist you've got it made."[6]

To see how accurate this characterization is, one can, for instance, look at the list of MacArthur fellowships awarded to particle theorists since the beginning of the fellowship program in 1981. There have been a total of nine such awards, all of them to string theorists (Daniel Friedan, David Gross, Juan Maldacena, John Schwarz, Nathan Seiberg, Stephen Shenker, Eva Silverstein, and Edward Witten), except for one, which went to Frank Wilczek in 1982.

Influence and power in an academic field is very much in the hands of those who hold tenured professorships at the highest-ranked universities. In the United States, if you believe *US News and World Report*, the top half-dozen physics departments are Berkeley, Caltech, Harvard, MIT, Princeton, and Stanford. The cohort of tenured professors in particle theory at these institutions who received their PhD after 1981 is a group of twenty-two people. Twenty of them specialize in superstring theory (a couple of these work on brane-worlds), one in phenomenology of supersymmetric extensions of the standard model, and one in high-temperature QCD.

The success that superstring theorists have had in fundraising and building institutions around the subject is also very impressive. The former head of the McKinsey management consulting firm has recently given $1 million to endow the Frederick W. Gluck chair in theoretical physics at the University of California at Santa Barbara, a chair now held by David Gross. The press release announcing this tells us that the donor was attracted by string theory:

> What brought Gluck and Gross together was string theory . . . Riveted by the subject, Gluck became a proselytizer for string theory by, for instance, giving his own presentation at Birnan Wood Golf Club.

A friend of Daniel Friedan's told me the following, perhaps exaggerated, story about how the superstring theory group at Rutgers came into being. Friedan was working in his office at the University of Chicago in the late 1980s when he got a phone call from one of the top officials of Rutgers University. This official asked Friedan what it would take to bring him and some other prominent superstring theorists to Rutgers. Since Friedan had little desire to move to New Jersey, he reeled off a list of what he considered to be completely outrageous demands: very high salaries, nonexistent teaching responsibilities, ability to hire at will a large number of more junior people and visiting faculty, a special building, and so on. The Rutgers official thanked him and hung up, leaving Friedan convinced he would never hear any more about this. A couple of hours later, though, he received a phone call informing him that Rutgers would gladly meet all his requests.

A superstring theorist looking for a pleasant place to spend a week or so at someone else's expense will in most years have a choice of thirty or so conferences, many in exotic locations. In 2002, for example, among the most prestigious and difficult to arrange options was a summer workshop in Aspen, but during the year, other possible destinations were Santa Barbara, Chile, Trieste, Genoa, the Black Sea, Corsica, Paris, Berlin, Vancouver, Seoul, China, and many others, including Baku in Azerbaijan.

As can be seen from this list, superstring theory's power and glory is not restricted to the United States, but extends throughout the world. Much of the leadership of the field is based in the United States, but the globalization phenomenon, which has made American culture such a dominant force around the globe, is also at work in this area.

A great deal of effort by superstring theorists has gone into publicizing the theory. My colleague at Columbia Brian Greene, a talented scientist, expositor, and speaker, has had two hugely successful books about superstrings, *The Elegant Universe*[7] and *The Fabric of the Cosmos.*[8] In 2003 there was a three-hour *Nova* television series based on the first of these books, and its $3.5 million cost was partly financed by the National Science Foundation. Superstrings and superstring researchers have been the subject of a large number of articles in the popular press, and until recently virtually all of them adopted an uncritical attitude toward the claims being made for the theory. The *New York Times* even went so far as to headline one of its articles on brane-worlds "Physicists Finally Find a Way to Test Superstring Theory," a claim that has little relation to reality.[9]

In the summer of 2002, the Institute for Advanced Study in Princeton organized a two-week summer program for graduate students to prepare them to become superstring theory researchers. This was the first of an annual series of such programs, with the one in 2003 covering both superstring theory and cosmology, the topic for 2004 again superstring theory, although for 2005 the program concerned physics at high-energy colliders. Superstring theory is being taught not just at the graduate level. The MIT physics department offers an undergraduate course in string theory, and as with all MIT courses, the course materials are available without cost online. A textbook for the course now exists, entitled *A First Course in String Theory.*[10] During 2001, the Institute for Theoretical Physics in Santa Barbara held a workshop for high-school teachers on superstring theory, evidently with the idea that this was something they should be teaching their students. There is no evidence the teachers were told much about the problems of the theory, and if one listens to the online proceedings, one can hear a high-school teacher saying that he has learned, "We may have to come up with new standards of what it means to say we know something in science."

We have seen that superstring theory is the only game in town. Why is this the case? What possibility is there for new ideas to come along and change the current situation? One common reaction I have received from physicists and mathematicians to this kind of question is the expression of a hope that somewhere, somehow, some young physicist is out there working on a new idea that will change everything. To understand the prospects for this happening, one needs to look carefully at what the standard career path is for ambitious, talented young physicists. This varies somewhat in different areas of the world, but I'll concentrate on the situation in the United States, both because I know it best and because of the leadership role played by American academics in the field.

The job situation for physicists in academia has been quite difficult since about 1970. Before that time, the American university system was expanding quickly, and the median age of tenured physics professors was under 40.[11] Students with PhDs in particle theory who wanted to find a permanent academic position could reasonably expect to be able to find one without too much trouble. After the 1970 recession, academic hiring never recovered, and from that time on, the average age of physics faculty began rising linearly for many years at a rate of about eight months per year. The latest figures show that the average age of tenured physics faculty members has now reached nearly 60.[12] The period of the last thirty years has thus been characterized by very little hiring of permanent physics faculty, while graduate programs have continued to turn out a large number of PhDs, making for grim job prospects for a young PhD in particle theory. It is also important to remember that before the early 1970s, quantum field theory was in eclipse, so the period during which most of the present particle theory professoriate was hired was one during which few people specialized in quantum field theory. This has been true again since 1984, as superstring theory came into vogue. It was only roughly during the decade 1974–1984 that quantum field theory was the area in which most new PhDs began their research careers. This was a decade during which an unusually small number of young theorists were able to get permanent jobs.

For the past several years, the Particle Data Group at Berkeley has collected data on particle theorists and particle experimentalists as well as on particles.[13] In each of the past few years their data show about 400–500 particle theory graduate students, and about 500 tenured faculty in particle theory. Since graduate students take about five years to get their degrees, one five-year cohort of students is nearly large enough to replace the entire tenured faculty in this field. A survey done in 1997 found an average of 78 students receiving PhDs in particle theory each year, 53 of these at the top 30 universities.[14]

Virtually all students completing PhDs who continue to do research in particle theory go next into a postdoctoral research position. These have set terms of from one to three years and are mostly funded by grants from the National Science Foundation (NSF) or Department of Energy (DOE). The 1997 survey and the more recent Particle Data Group survey both found about 200 theorists holding postdoctoral research positions. It is very common for recent PhDs to hold a sequence of postdoctoral positions, often at different institutions. These positions cannot be held indefinitely, so sooner or later, to continue being able to do research, one needs to find a tenure-track academic job at an institution that supports this, often meaning one with a graduate program. The situation with these can be seen in detail at a website called the "Theoretical Particle Physics Jobs Rumor Mill,"[15] which keeps close track of which jobs are available, who gets on which short list for hiring, and who is ultimately hired. Looking over the data from the last few years, on average, about 15 theorists are hired into these tenure-track positions each year.

In principle, someone who takes such a tenure-track position will be considered for tenure about six years later, although this will often be done earlier for someone who is doing very well and in danger of being hired away by another institution. I know of no data about how many of the 15 theorists who get tenure-track jobs ultimately end up in a permanent tenured position, but it is probably roughly around 10. Some people do not get tenure, some leave the field for other reasons, and people move from one such job to another, so there is a certain amount of double counting in the average number of 15.

Thus, the bottom line is that of about 80 students getting PhDs in particle theory each year during recent years, perhaps 10 of them can expect ultimately to have a permanent position doing particle theory research. What happens to those who lose out in this academic game of musical chairs? Typically, they are faced with the frightening task of starting a completely new career, but most end up doing well. Some find academic jobs at colleges where teaching rather than research is emphasized, some go to law school or medical school, and in recent years many have gone to work in the computer or financial industries.

Some of them do very well indeed, including several of my roommates from graduate school, whose PhDs were in quantum gravity and particle theory (Nathan Myhrvold and Chuck Whitmer). After short stints as theoretical physics postdocs (Nathan worked in Cambridge with Stephen Hawking), they started up a computer software company called Dynamical Systems near Berkeley and occasionally asked me to join them. Since I was being paid a reasonable sum as a postdoc at Stony Brook to work on whatever I wanted, I didn't find the idea very appealing of going to work with them and spending long hours writing computer code in return for stock that seemed likely to end up being worthless. This turns out to have been a big mistake, as Nathan sometimes reminds me when he comes to New York in his personal jet. Dynamical Systems stock turned out to be quite valuable when the company was bought early on by Microsoft. Nathan, Chuck, and some of the others at Dynamical Systems went to work for Microsoft, with Nathan ultimately becoming the company's chief technology officer.

While particle theorists who do not get one of the few permanent academic jobs end up doing many different things, there is one thing they rarely end up doing: particle theory. The days of Einstein being able to do important work during his spare time while working at the Patent Office are long gone, victim of both the greatly increased complexity and sophistication of theoretical physics and of the increased demands of time and energy in many professions. It is an unfortunate fact that new advances in particle theory are unlikely to come from anyone who is not either being paid to think about the subject or independently wealthy.

How does one win at this game and get a permanent academic job? The rules are quite straightforward and well understood by everyone involved. Starting in the year in which one gets one's PhD, one needs to overcome a very specific hurdle every couple of years: that of convincing a hiring committee of senior theorists at some institution to choose one from among a large number of applicants. Many of one's competitors' files will have letters from prominent physicists and a sizable number of published papers, some perhaps even on the latest and hottest topic. One's papers had better have been accepted for publication by referees at some of the best journals, and should be on topics that one's evaluators will recognize as being of significance and importance.

If one wins this competition and gets the job, with bad luck it will be a one-year position and one will have to be sending out new applications within a few months after one arrives to take it. More likely, one will have a year or perhaps even two after the start of the new job to prepare for the next hurdle by getting new research done, and as many papers as possible accepted for publication. When one starts a new research project there are difficult choices to be made. Should one work on a certainty, such as a small advance related to what one has done before? What about trying to master the latest hottest topic in the field, seeing whether one can find some aspect of it on which no one else has yet published and finish work on it before anyone else? Perhaps one should try to work out some unusual idea that seems promising but that no one else seems to find interesting? Before deciding on the latter, one needs to worry about whether there is a good reason no one else is working on this idea, a reason that one may not figure out for a year or so, at which point one is in danger of having to go empty-handed before the next round of hiring committees.

Isadore Singer is a prominent mathematician who has worked for many years on problems at the interface of theoretical physics and mathematics. He made the following comments in an interview that took place in 2004 at the occasion of his being awarded the Abel

Prize, which he shared with Michael Atiyah for their work on the Atiyah–Singer index theorem:

> In the United States I observe a trend toward early specialization driven by economic considerations. You must show early promise to get good letters of recommendations to get good first jobs. You can't afford to branch out until you have established yourself and have a secure position. The realities of life force a narrowness in perspective that is not inherent to mathematics . . . When I was young the job market was good. It was important to be at a major university, but you could still prosper at a smaller one. I am distressed by the coercive effect of today's job market. Young mathematicians should have the freedom of choice we had when we were young.[16]

Singer's comments apply to young mathematicians who would like to branch out into new mathematics related to physics, but even more to young particle theorists, for whom the job market is even more competitive than it is in mathematics.

What happens to those who successfully make it through this system and finally arrive at the holy grail of a tenured academic position? They are now one of a small number of people responsible for the survival of the theory group in the physics department at their institution. This group probably has a grant from either the DOE or the NSF. In fiscal year 2001, the DOE spent about $20 million funding about 70 such groups, including 222 faculty, 110 postdocs, and 116 graduate students. The NSF spends roughly half as much as the DOE, supporting about half as many people. A typical DOE or NSF grant at a major institution will be renewable every five years and provide half a million dollars or so per year in support. A large chunk of this will go to "overhead," basically a payment to the university that is supposed to cover costs for the physical plant, libraries, etc. used by the theory group. This kind of grant income is crucial to most university budgets, and underwrites their willingness to provide well for the faculty members on the grant, for instance by keeping their teaching loads low. Some of the rest of the grant goes toward paying the salaries of a couple of postdocs, and some goes for research fellowships for a few graduate students working in the theory group. The research fellowships pay a stipend to the student and tuition to the university. Without the tuition income from these fellowships, universities would be highly likely to cut back on the number of theoretical physics graduate students they enroll.

Much of the grant goes into direct payments to the faculty members involved called "summer salary." These are based on the philosophy that what the university pays professors covers only nine months of their time, so they may accept pay from elsewhere for up to three months per year. The NSF and DOE pay particle theorists up to two-ninths of their university salary for the nominal reason that otherwise they might be finding other summer employment or teaching summer school, although in practice neither of these is actually very likely. Finally, grants also cover other things such as travel expenses to conferences and costs of office computers, although these are typically much smaller than the costs of salaries.

The overall dollar amount of DOE and NSF funding of theoretical physics has not changed a great deal during the past decade. Since salaries have increased considerably, the number of researchers supported by these grants has fallen significantly. Each time one's grant is up for renewal, the danger of having it cut back in some way or, even worse, canceled completely, is very real. The loss of a grant could mean losing two-ninths of one's income, being unable to hire postdocs or support graduate students, lacking money for travel to conferences, and also the possibility that the university would start looking into increasing one's teaching load. These are consequences most theorists would like to avoid at all costs, so tenured faculty members are again in the situation of having periodically to beat out the competition before panels of much the same composition as the hiring committees they dealt with before.

A British superstring theorist, Michael Duff, who has thrived in the American system (although in 2005 he returned to the UK), contrasts it to that in Britain as follows:

> Competition, even—or perhaps, especially—in academia is cut-throat and the British notion of "fair play" does not apply.
>
> I hope my American friends will not be offended when I say that ethical standards are lower as a consequence. The pressure to succeed

is exacerbated for university academics who are traditionally paid for only nine months of the year and must seek research funding from agencies such as the National Science Foundation for their summer salaries. An incredible amount of time and effort is thus spent preparing grant proposals.[17]

The particle theory community in the United States is not a very large one, consisting of a total of about a thousand people. It is a very talented group, but has now been working for two decades in an environment of intellectual failure and fierce competition for scarce resources. There are other reasons why there is only one game in town, but the social and financial structures within which people are working are an important part of this situation.

The Landscape of String Theory

The last few years have seen a dramatic split in the ranks of superstring theorists over something called the anthropic principle. The anthropic principle comes in various versions, but they all involve the fact that the laws of physics must be of a nature that allows the development of intelligent beings such as ourselves. Many scientists believe that this is nothing more than a tautology, which while true, can never be used to create a falsifiable prediction, and thus cannot be part of scientific reasoning. Controversy has arisen as a significant group of superstring theorists have begun to argue that superstring theory's inability to make predictions is not a problem with the theory, but a reflection of the true nature of the universe. In their view, the lesson of superstring theory is that predicting many if not all of the parameters that determine the standard model is inherently impossible, with only the anthropic principle available for explaining many aspects of why the universe is the way it is.

Recall that superstring theory suffers from the following vacuum degeneracy problem. Since one doesn't know what the underlying fundamental M-theory is, superstring theorists take the first terms in the perturbative string theory expansion in the number of holes of the world sheet of the string. They assume that this calculation will give something fairly close to what one would get from a calculation in the true M-theory. To set up this approximate calculation, one has to choose a background ten- or eleven-dimensional space-time, and perhaps also some choices of "branes," i.e., certain subspaces of the full space-time to which ends of strings are attached. Such a choice is referred to as a choice of vacuum state, since the hope is that it corresponds to the choice of a lowest-energy state in the unknown M-theory. There are many, perhaps infinitely many, classes of background spaces that appear to be possible consistent choices, and each one of these classes comes with a large number of parameters that determine the size and shape of the background space-time. These parameters are known as moduli, since historically a modulus function is one whose values can be used to parameterize the size or shape of a space.

The hope has been that the values of these moduli are somehow determined by the unknown dynamics of M-theory. To do this, some mechanism has to be found that gives different energies to vacuum states corresponding to different values of the moduli. If the energy of the vacuum states does not depend on the moduli, one expects on general principles that the moduli will give rise to quantum fields corresponding to massless particles, and these have not been observed. The picture of an energy function depending on many moduli parameters has come to be known as the "landscape" of superstring theory. This terminology comes from taking the altitude in a landscape to be the analogue of the energy, and the latitude and longitude of a point in the landscape to be the analogues of two moduli parameters.

In 2003, a mechanism was found by the physicists Kachru, Kalloh, Linde, and Trivedi that can potentially give different energies for different values of the moduli in such a way as to allow one to fix their values by finding minima of the energy as a function of the moduli.[1] In the landscape picture, these minima are the bottoms of various valleys. This "KKLT" mechanism is quite complicated, so much so that Shamit Kachru's Stanford colleague Leonard Susskind refers to it as a "Rube Goldberg machine," with Kachru the "master Rube Goldberg architect."[2] Starting with a Calabi–Yau space to compactify six of the ten dimensions of a background for superstring theory, KKLT adds several extra layers of structure involving branes and fluxes. These fluxes are generalizations of magnetic fields to higher dimensions, with the fields trapped by the topology of the Calabi–Yau space.

The KKLT mechanism picks out not a unique value for the values of the moduli, but a very large set of values, any one of which should be as good as any other. Estimates of the number of these possible values are absurdly large (e.g., 10^{100}, 10^{500}, or even 10^{1000}), far beyond the number of particles in the universe or anything else one can imagine counting. While string theory is supposed to be a Theory of Everything, Kachru refers to this elaboration of it as a Theory of More Than Everything. The consistency of the KKLT mechanism is still debated by superstring theorists, a debate that may never be resolved, since one doesn't know what underlying M-theory governs this situation.

The possible existence of, say, 10^{500} consistent different vacuum states for superstring theory probably destroys the hope of using the theory to predict anything. If one chooses among this large set just those states whose properties agree with present experimental observations, it is likely there still will be such a large number of these that one can get just about whatever value one wants for the results of any new observation. If this is the case, the theory can never predict anything and can never be falsified. This is sometimes known as the "Alice's Restaurant problem," from Arlo Guthrie's famous refrain, "You can get anything you want at Alice's Restaurant." Being able to get anything one wants may be desirable in a restaurant, but isn't at all in a physical theory.

In recent years, Susskind, one of the codiscoverers of string theory, has begun to argue that this ability of the theory to be consistent with just about anything should actually be thought of as a virtue. He argues that the cosmological constant in the different states will take on a discrete but nearly continuous set of values (some call these possible values for the cosmological constant the "discretuum"). Recall that combining supersymmetry and gravity predicts that the energy scale for the cosmological constant is at least 10^{56} times larger than its observed value. It can be argued that the existence of the discretuum implies that at least some possible vacuum states of superstring theory will have unusually small cosmological constants, so small as to be in agreement with experiment. Susskind takes the point of view that the existence of huge numbers of possible vacuum states in superstring theory is actually a virtue, because it allows the possibility of the cosmological constant being small enough in at least some of them.

In 1987, Steven Weinberg published an article arguing that in order for galaxies to form and life as we know it to develop, the cosmological constant could not be too large.[3] If it were more than ten to one hundred times larger than what seems to be its value, the universe would have expanded too rapidly for galaxies to be produced. Weinberg suggested that perhaps the explanation of the problem of the small size of the cosmological constant was the anthropic principle. The idea is that there are a huge number of consistent possible universes, and that our universe is part of some larger multiverse or megaverse. Quite naturally, we find ourselves in a part of this multiverse in which galaxies can be produced and thus intelligent life can evolve. If this is the case, there is no hope of ever predicting the value of the cosmological constant, since all one can do is note the tautology that it has a value consistent with one's existence.

Susskind has recently been campaigning vigorously among the particle theory community for his point of view, stating,

> Ed Witten dislikes this idea intensely, but I'm told he's very nervous that it might be right. He's not happy about it, but I think he knows that things are going in that direction. Joe Polchinski, who is one of the really great physicists in the world, was one of the people who started this idea. In the context of string theory he was one of the first to realize that all this diversity was there, and he's fully on board. Everybody at Stanford is going in this direction.[4]

In a February 2005 Stanford University press release, Susskind describes the various possible vacuum states of string theory as "pocket universes" and insists that the arguments for their huge number are correct, writing, "Ed Witten worked very hard to show that there was only a very small number, and he failed—failed completely."[5] He goes on to claim, "More and more as time goes on, the opponents of the idea admit that they are simply in a state of depression and desperation." Late in 2005, his popular book *The Cosmic Landscape: String Theory and the Illusion of Intelligent Design*[6] appeared, promoting his point of view on string theory to a wide audience.

Witten has been quoted in a *New York Times* article as saying, "I continue to hope that we are overlooking or misunderstanding something and that there is ultimately a more unique answer."[7] At a talk at the Kavli Institute for Theoretical Physics in October 2004 on "The Future of String Theory,"[8] he said, "I'd be happy if it is not right, but there are serious arguments for it, and I don't have any serious arguments against it."

As we saw in the first chapter of this book, David Gross has expressed a much more full-throated disapproval of the anthropic principle, invoking Einstein's core beliefs and Churchill's purported advice to "Never, never, never, never, never give up." Both Witten and Gross continue to hope that somehow the implications of the possible existence of a huge number of consistent backgrounds for superstring theory can be overcome. Gross believes that the conclusion that superstring theory cannot explain fundamental features of our universe is premature, stating,

> We still do not know in a deep sense what string theory is. We do not have a fundamental, background independent, formulation of the theory. We may have 10^{1000} consistent metastable vacua, but not a single consistent cosmology. Perhaps there is a unique cosmology.
>
> Many string theorists suspect that a profound conceptual change in our concept of space and time will be required for the final formulation of string theory. If so, the criterion for determining the state of nature (the vacuum) could be very different. There is no reason, at this preliminary stage of our understanding, to renounce the hope that it will yield a truly predictive theory of the universe.[9]

Perhaps the unknown M-theory built on new concepts of space and time that Gross hopes for really exists and has a unique vacuum that explains the properties of the universe, but more and more string theorists now believe that this is little more than wishful thinking. Gross's colleague Polchinski wrote the following about the attitudes of Gross and Witten:

> In fact in string theory there is a cult of "monovacuism," whose prophet resides in New Jersey (or possibly in the office below mine), to the effect that some magic principle will pick out a single vacuum, namely ours. I would like this to be true, but scientists are supposed to be immune to believing something just because it makes them happy.[10]

A CERN string theorist, Wolfgang Lerche, went so far as to claim that the existence of such a vast number of possible string theory vacuum states was obvious long before the 2003 KKLT work:

> Well, what I find irritating is that these ideas are out since the mid-80s; in one paper on 4d string constructions a crude estimate of the minimal number of string vacua was made, to the order 10^{1500}; this work had been ignored (because it didn't fit into the philosophy at the time) by the same people who now re-"invent" the landscape, appear in journals in this context and even seem to write books about it . . . the whole discussion could (and in fact should) have taken place in 1986/87. The main thing that has changed since then is the mind of certain people, and what you now see is the Stanford propaganda machine working at its fullest.[11]

As more and more superstring theorists have come to the conclusion that superstring theory really does have all these vacuum states, and inherently cannot predict the cosmological constant and possibly other undetermined parameters of the standard model, one often hears the following analogy. In 1596 Kepler proposed a mathematically elegant conjectural explanation for the distances between the orbits of the six known planets, an explanation that invoked the fact that there are only five Platonic solids. Of course, it later became clear that the distances between the planets are the product of a number of historical contingencies in the evolution of the solar system, and not the sort of thing that fundamental physical laws can predict. The argument is that perhaps many, if not all, of the aspects of the standard model for which we have no explanation are actually just environmental, depending on the particular state of the universe we find ourselves in, and not the direct manifestation of any underlying physical laws. If this is the case, then the only possible predictions of these things are those coming from the anthropic constraint of making our existence possible.

The problem with this argument is that in the case of the solar system, the relevant physical theory (Newtonian mechanics) comes with a well-defined understanding of which things are determined by the underlying theory, and which are environmental accidents of history. Superstring theory comes with no such distinction. No one knows how to determine what the theory is capable of predicting, and what can't be predicted because it is environmental. There does not seem to be any aspect of the standard model that superstring theorists are sure can be predicted in principle by the theory.

A group of theorists, including Michael Douglas, of Rutgers, have claimed that one can hope to make predictions from superstring theory by analyzing the statistics of the possible vacuum states consistent with our existence. If the great majority of these states have some given property, then they expect that we will see this property in our particular universe. The simplest such property that they have tried to analyze is that of the energy scale of supersymmetry breaking. Are states with supersymmetry breaking at a very high energy scale, say the Planck scale, more common than states where this happens at a low-energy scale, one that would be accessible to observation at the Large Hadron Collider (LHC) at CERN?

It's very unclear that this question can sensibly be addressed. First of all, calculating what happens in each of the 10^{500} or more conjectural states seems likely to be an impossible task. Even if this could be done, one would have no idea what probability to assign to each of these states, since the likelihood of their occurrence depends on details of the dynamics of the big bang that are not understood at all. One can just assume equal probability for every state, but even this fails to lead anywhere if the number of states is infinite, which appears likely to be the case. This problem can be dealt with by putting in some sort of cutoff to make the number of states finite, but then results depend on the choice of cutoff.

Douglas and others still hoped to be able to make some sort of prediction and wrote several papers on the subject. The story of these papers is a remarkably confused one. One of Douglas's papers on the subject[12] ended up being posted in four different versions, with significant changes in its conclusions between versions. Based on one of the earlier versions of the Douglas paper, Susskind posted a paper[13] that claimed that some of his own earlier arguments on the subject were wrong. When Douglas changed his conclusions, Susskind withdrew the paper, presumably because its claims that his own earlier paper was wrong were now themselves wrong, because they had been based on Douglas's now incorrect paper. Both this withdrawn paper and the earlier one to which it referred are unusually short and lacking in anything like careful mathematical arguments.

A bizarre episode involving Susskind took place in late July 2004, beginning with the appearance of a paper by the physicist Lee Smolin explaining in detail why the anthropic principle could never yield falsifiable predictions, and thus did not deserve to be thought of as a scientific principle. Susskind the next day tried to post to the preprint archive a three-page paper, half of which consisted of a reproduction of a letter from Smolin outlining his argument, with the other half an attack on Smolin in which he acknowledged that he had not carefully read Smolin's paper. This paper was rejected by the administrators of the archive, something that is extremely unusual. The archive is intended as a repository of unrefereed work, and I know of no other example of its administrators rejecting a paper by a well-known, mainstream physicist.

Around the same time, the Russian physicist and string theorist Alexander Polyakov (now at Princeton) posted an article reviewing his career and his efforts to understand the duality relation between gauge theory and string theory. In this article he states,

> In my opinion, string theory in general may be too ambitious. We know too little about string dynamics to attack the fundamental questions of the "right" vacua, hierarchies, to choose between anthropic and misanthropic principles, etc. The lack of control from the experiment makes going astray almost inevitable. I hope that gauge/string duality somewhat improves the situation . . . Perhaps it will help to restore the mental health of string theory.[14]

In his recent book, Susskind admits that he has no plausible idea about how one might be able to derive any predictions from string theory. The surprising thing is that he and other prominent theorists don't see this as a reason to give up on the theory, but instead choose to believe that the theory must be true, even though it can't predict anything. Susskind refers to objections that string theory is not falsifiable as "pontification, by the 'Popperazi,' about what is and what is not science," and goes on to write, "I am inclined to think that no idea can have great merit unless it has drawn this criticism."[15]

One would expect that once theorists could no longer see a way forward to use a theory to make predictions, they would abandon it and work on something more promising. This does not seem to be happening. One question surrounding superstring theory has always been that of what it would take to convince its adherents that it is an unworkable idea. Since there is no well-defined theory, the theory can never be shown to be wrong in the sense of leading to contradictions or predictions that disagree with experiment. One might think that the only hope of showing that the theory had failed would be by demonstrating that it is vacuous and could never predict anything, but it appears that even that is not enough to change minds.

Princeton cosmologist Paul Steinhardt believes that it is not the opponents of the landscape scenario who are desperate, but rather those string theorists like Susskind who have turned to the anthropic principle. He says,

> String theorists have turned to the anthropic principle for salvation.
>
> Frankly, I view this as an act of desperation. I don't have much patience for the anthropic principle. I think the concept is, at heart, nonscientific. A proper scientific theory is based on testable assumptions and is judged by its predictive power. The anthropic principle makes an enormous number of assumptions regarding the existence of multiple universes, a random creation process, probability distributions that determine the likelihood of different features, etc. none of which are testable because they entail hypothetical regions of spacetime that are forever beyond the reach of observation. As for predictions, there are very few, if any. In the case of string theory, the principle is invoked only to explain known observations, not to predict new ones. (In other versions of the anthropic principle where predictions are made, the predictions have proven to be wrong. Some physicists cite the recent evidence for a cosmological constant as having been anticipated by anthropic argument; however, the observed value does not agree with the anthropically predicted value.)[16]

He goes on to refer to the "current anthropic craze" as "millennial madness."

The anthropic argument that Susskind and others take refuge in is actually a red herring. Gross describes the situation thus: "We see this kind of thing happen over and over again as a reaction to difficult problems . . . Come up with a grand principle that explains why you're unable to solve the problem."[17]

Even if one believes that the cosmological constant can never be fixed by theory and can only be anthropically determined, this has little to do with the problems faced by superstring theory. If the theory made some accurate predictions, but left the cosmological constant undetermined, one might take the anthropic argument seriously. Instead, the fact of the matter is that the theory not only doesn't predict the cosmological constant, it doesn't predict anything at all. Whether or not anthropic reasoning ever turns out to be necessary in physics, in this case it is nothing more than an excuse for failure. Speculative scientific ideas fail not just when they make incorrect predictions, but also when they turn out to be vacuous and incapable of predicting anything.

After talking to several string theorists about the current situation, a reporter for *Science* magazine wrote, "Most researchers believe that a huge number of distinct versions of the theory may jibe with what we know and can measure. If so, physicists may have to rethink what it means for a theory to explain experimental data."[18]

This is utter nonsense. There is no need at all to rethink what it means for a theory to explain experimental results. What has happened is that in order to avoid admitting failure, some physicists have tried to turn Feynman's comment, "String theorists make excuses, not predictions," from a criticism into a new way of pursuing theoretical science.

Other Points of View

This book has surveyed the current state of fundamental particle physics from a very particular point of view, that of a mathematically minded particle physicist. The emphasis has been on the standard model, the mathematics behind it, and the accelerator-based experimental techniques that led to its discovery and whose limitations now make further progress difficult. There are other points of view on the problems of particle physics, and this chapter will consider some of them.

After taking account of the limitations of particle accelerator technology, one obvious question is whether there is some other way of studying high-energy particle interactions. Some of the earliest discoveries in particle physics were made not with accelerators, but by studying cosmic rays. Elementary particles and nuclei accelerated by various astrophysical processes are continually raining down on the earth, and colliding with other particles in the atmosphere. One can analyze the products of these collisions to see whether new particles have been produced and to check our understanding of high-energy particle collisions. To produce a center-of-mass energy equivalent to that of the LHC (14 TeV = 1.4×10^{13} eV) would require a cosmic ray of energy about 10^{17} eV, since it would be hitting an essentially fixed target. Cosmic rays at this energy and above have been observed, but their number is quite small, with only a few per century hitting each square meter of the earth's surface. Recall that as the energy of collisions increases, the probability of an interesting collision, one in which a large quantity of energy is exchanged and new particles are produced, goes down rapidly. This is why extremely high luminosities are required at accelerators such as the LHC in order to be able to get interesting results.

Collisions with center-of-mass energy ten times that of the LHC occur about a hundred times a year over an area of a square kilometer. In

2005, the Auger observatory (named after Pierre Auger, who made early observations of cosmic rays) began operation. It has detectors covering 3000 square kilometers in Argentina, and is designed to study the highest-energy cosmic rays that have ever been observed, those with energies of about 10^{20} eV. While the Auger observatory will not be able to say much about the interactions of particles at this energy, the very fact of the existence of such particles is of interest. At these high energies, particles traveling through interstellar space scatter off the low-energy photons making up the so-called cosmic microwave background. This scattering causes them to lose energy, so one expects to see few if any particles with energies high enough to be subject to this scattering effect. If Auger does see such particles, it will be evidence for some new and not yet understood physics.

The biggest particle accelerator of all is the big bang, so it is natural that in recent years many particle theorists have turned to trying to look at particle physics from the point of view of cosmology. This is a large topic that would require another book to address, and this author is not competent to write it, but it is so important that at least a few comments are in order. Very soon after the development of the standard model in the early 1970s, some particle theorists turned to the problem of trying to use it to model the big bang. Steven Weinberg's 1977 book *The First Three Minutes: A Modern View of the Origin of the Universe*[1] gives an excellent popular account of this early work and should be consulted for more details.

Modern cosmological theory implies that the universe was increasingly hot and increasingly dense at moments closer and closer to that of the big bang. High temperatures imply high energies of the particles involved, so the hope has been that if one can see effects from the very earliest moments after the big bang, these might tell us something about the behavior of particles at these energies. Unfortunately, one can't see back that far in time, but can see only how it all turned out. By extrapolating back from the present, it is possible to estimate the primordial abundance of various elements and determine that the early universe was mostly hydrogen and helium. Early universe models based on the particle physics of the standard model are able to reproduce the observed abundances of these elements. A question that still remains open is that of baryogenesis: why is the matter in our universe mostly baryons (protons and neutrons), with hardly any antibaryons (antiprotons or antineutrons)? At high enough temperatures a more or less equal number of baryons and antibaryons should have been created, and one needs to explain the asymmetry that left a certain number of baryons around after most of the baryons and antibaryons had annihilated each other. There are various possible sources for this asymmetry, but it remains unknown how it came about.

In the late 1970s and early 1980s there was much optimism that cosmology would give particle physics some information about the physics occurring at GUT (grand unified theory) energy scales, but unfortunately this so far has not really worked out. Perhaps the answer to the question of baryogenesis lies in GUT-scale physics, but an unambiguous resolution of this question remains elusive. Attention has turned away from the earliest, hottest times to later ones, for which we have much more experimental data.

The cosmic microwave background (CMB) is radiation discovered by Arno Penzias and Robert Wilson in 1965. It is blackbody radiation of cosmological origin at the very low temperature of 2.7 degrees Kelvin. According to current theory, this radiation consists of leftover photons from a time 400,000 years after the big bang, the time at which electrons and protons stopped existing in a hot plasma of free particles and combined to form electrically neutral atoms. Before this time, photons would have been continually scattering off the plasma of charged particles, but after this time they could travel without interference. The CMB radiation we see today consists of these remnant photons, still containing information about their origin at this relatively early point in the universe's history.

At the time the CMB radiation was produced, the temperature of the universe was about 3,000 degrees Kelvin. This is very hot by normal standards, but corresponds to particle energies of only a few tenths of an electron volt. So while it is not possible to use the CMB radiation to see directly the effects of very high energy particle interactions, there is still a huge amount of information contained in this radiation. Until recently, observations of the CMB just showed structureless blackbody radiation, the same in all directions. In 1992, the

COBE (Cosmic Background Explorer) satellite experiment was first able to observe some anisotropies, or structure, in the CMB. By 2003 a more sophisticated satellite experiment, WMAP (the Wilkinson Microwave Anisotropy Probe), was able to report the first results of a more detailed look at this structure, gathering a wealth of new information about the early universe. New data continue to come in from the WMAP satellite, and a next-generation satellite, called Planck, is to be launched in 2007.

Astrophysicists continue to try to extract more information from the WMAP CMB anisotropy data and look forward to the data from Planck. One hope is that the Planck satellite will be able to see the effects of early-universe gravitational waves on the polarization of the CMB radiation. Unlike photons, such gravitational waves would not be scattered by a charged plasma, so if their effects are seen, one potentially would have a window into the very early universe, possibly even into the conjectured phase of exponential expansion predicted by inflationary cosmological models.

Using the WMAP data and other astronomical observations, in particular those of supernovae at great distances, cosmologists have constructed a "standard model" of the universe, one that raises two puzzles for particle physicists. In this model only five percent of the energy density of the universe is contained in normal matter made out of baryons. Twenty-five percent is contained in cold dark matter, whose nature remains unknown. It is possible that cold dark matter consists of a new type of stable particle, one that would have neither electric charge nor strong interactions, but would have astrophysical effects through purely gravitational interactions. Such conjectural particles include so-called WIMPs (weakly interacting massive particles), and one argument given for supersymmetric versions of the standard model is that they can contain such a stable particle.

Finally, 70 percent of the energy density of the universe seems to consist of dark energy, a uniform energy density of the vacuum (the cosmological constant). We saw earlier in this book how supersymmetric models of particle physics run into trouble because they require that the vacuum state break supersymmetry, and as a result must have an energy density many, many orders of magnitude larger than this observed number. The landscape argument that the number of possible vacuum states is so huge that there should be some in which this number is much smaller than expected (and thus in agreement with experiment) was mentioned in the previous chapter.

While cosmology has yet to resolve any of the outstanding problems of the standard model, the two new puzzles it has provided may be important clues. Is there a new unknown stable particle that makes up the cold dark matter? What is the origin of the vacuum energy density and how can one calculate it? Both questions still remain completely unresolved.

As we have seen, the question of how to construct a quantum version of general relativity, Einstein's theory of the gravitational force, still remains open. One of the main motivations of superstring theory is to provide such a construction, but we have seen that it has not yet completely successfully done so. The mathematical structure of general relativity is tantalizingly close to that of the standard model, since it is fundamentally a geometric theory. From a geometer's point of view, the Yang–Mills gauge fields are connections telling one how to compare fields at neighboring points. General relativity can also be expressed in terms of such connections, which in this case describe how to compare vectors at neighboring points. But the geometry behind general relativity, Riemannian geometry, contains extra structure that does not occur in the Yang–Mills case. This extra structure is that of a metric, i.e., a way of measuring the sizes of vectors, and these metric variables require a different sort of dynamics from that of the Yang–Mills gauge fields. At long distances or low energies, this dynamic is known: it is determined by Einstein's field equations. If one tries to use this same dynamic in a quantum field theory at short distances, one runs into problems with infinities that cannot be dealt with by the standard renormalization methods.

String theory tries to deal with this problem by assuming that at short distance the fundamental fields of the theory are something nongeometrical: the excitation modes of a string. Another very different approach to the quantum gravity problem that has become popular in recent years goes under the name of loop quantum gravity (LQG). To describe the LQG program is a long story; for a popular version, see Lee Smolin's *Three Roads to Quantum Gravity*,[2] or for a more technical discussion see Carlo Rovelli's *Quantum Gravity*.[3] LQG uses the standard geometrical connection variables of general relativity in its quantization of gravity, but with nonperturbative quantization methods that differ from the standard Feynman diagram expansion that is known to run into problems with infinities.

While the LQG program has arguably been as successful as the superstring program in coming up with a plausible approach to the construction of a quantum theory of gravity, unlike superstring theory, LQG does not purport to give an explanation of the standard model. It is purely a theory of quantum gravity, one that in principle is independent of one's theory of the other particle interactions. Much of the interest in superstring theory originally came from the hope that it would provide not just a theory of gravity, but a unified theory of all particle interactions. In recent years, as it has become increasingly clear that this hope is a mirage, LQG has been drawing an increasing amount of attention, and an often hostile debate between partisans of these two different research programs has ensued. Many papers in string theory continue to start off with the motivational statement that string theory is the "most promising theory of quantum gravity" or something similar, a claim that infuriates physicists working on LQG. Superstring theory continues to draw by far the lion's share of resources, since its partisans dominate the field's major research centers, especially in the United States. While many different physics departments have active string theory groups and are willing to hire young string theorists, only a small number of institutions in the United States would even consider the idea of offering a job to a young physicist working on LQG.

Besides LQG, an even more speculative approach to quantum gravity that has come out of the study of general relativity is something known as twistor theory. This is an idea pioneered by Sir Roger Penrose together with many collaborators at Oxford and elsewhere. His remarkable book *The Road to Reality*[4] gives an extensive overview of theoretical physics, largely from the point of view of general relativity rather than particle physics, and can be consulted for a summary of the main ideas of twistor theory. Penrose himself takes the point of view that a successful theory of quantum gravity will not only involve twistor theory, but also require a reworking of the fundamental ideas of quantum mechanics, although he is very much in the minority in thinking this. Twistor theory involves ideas about geometry that are special to four dimensions, and makes fundamental use of the geometrical aspects of spinors and of complex geometry. It still has not led to a complete theory of quantum gravity, and like LQG, it does not purport to offer an explanation of the standard model.

Besides its potential applications to gravity, twistor theory has turned out to be quite useful in many other contexts, leading to exact solutions of several geometrically interesting systems of equations. These include the self-duality equations in Yang–Mills theory that turned out to be so important for the work of Donaldson on four-dimensional topology. Twistor theory methods also led to new formulae for certain scattering amplitudes in four-dimensional Yang–Mills quantum field theory, a fact that recently motivated Witten to try to express these amplitudes in terms of a topological theory of strings, where the strings live not in physical space-time, but in the space of twistors. While this has led to interesting new ways of calculating scattering amplitudes, it has not yet led to the hoped for equivalence between Yang–Mills quantum field theory and a new kind of string theory.

One other speculative research program that deserves mention goes under the name of noncommutative geometry and has been promoted by the French mathematician and Fields medalist Alain Connes. As noted in an earlier chapter, an algebra is essentially just an abstract mathematical structure whose elements can be consistently multiplied and added together, and the study of these is part of the subfield of mathematics that mathematicians also call algebra. There is a deep and fundamental link between the mathematical subfields of geometry and algebra. This link associates to a geometrical space a specific algebra: the algebra of functions defined on that space. This algebra of functions is commutative, i.e., when you multiply such functions it doesn't matter in which order you do the multiplication. Connes has pioneered the idea of studying more general noncommutative algebras by thinking of them as algebras of functions on a generalized kind of geometrical space, and this is what noncommutative geometry is about. He has some speculative ideas about how to understand the standard model using these notions of noncommutative geometry. For the details of this and to see how some of these new ideas about geometry work, one can consult Connes's research monograph on the subject.[5]

Conclusion

But to live outside the law, you must be honest.
—Bob Dylan, "Absolutely Sweet Marie"

The search for an understanding of the most fundamental objects in nature and how they interact to make up the physical world has a long and illustrious history, culminating in a truly fantastic success during the past century. The discovery of the standard model is an intellectual achievement that will be remembered for the rest of human history. One unexpected result of this progress has been that the field of theoretical particle physics has now been a victim of its own success for nearly a quarter century. Without any new experimental data to provide clues as to which direction to take in order to make further progress, the field has stagnated and worked itself a long way down a blind alley. In the past, when faddishness was driven by the need to understand new and unexplained experimental results, it was an efficient way of making progress, but the lack of useful experimental input has made the traditional organizational system of particle theory seriously dysfunctional.

Changes are desperately needed for particle theorists to find a way to live usefully outside the iron laws that used to be provided by direct contact with experiment, and one change that needs to occur is a dramatic increase in the honesty with which the results of theoretical speculation are evaluated. When new experimental results could be relied on to appear sooner or later and keep theorists honest, it was not so important that theorists themselves found ways to evaluate whether ideas were working out the way they were supposed to. Daniel Friedan made the following points, in the context of a recent discussion of the failure of superstring theory:

Recognizing failure is a useful part of the scientific strategy. Only when failure is recognized can dead ends be abandoned and useable pieces of failed programs be recycled. Aside from possible utility, there is a responsibility to recognize failure. Recognizing failure is an essential part of the scientific ethos. Complete scientific failure must be recognized eventually.[1]

The failure of the superstring theory program must be recognized and lessons learned from this failure before there can be much hope of moving forward. As long as the leadership of the particle theory community refuses to face up to what has happened and continues to train young theorists to work on a failed project, there is little likelihood of new ideas finding fertile ground in which to grow. Without a dramatic change in the way theorists choose what topics to address, they will continue to be as unproductive as they have been for two decades, waiting for some new experimental result finally to arrive.

The small hope that a dramatic increase in luminosity would turn up something at the Tevatron is now dwindling as the difficulty in achieving this becomes clear, so the possibility of new experimental discoveries will most likely have to wait at least until 2008, when results begin to come in from the LHC at CERN. Perhaps at that point the LHC will be able to answer questions about the origin of the electroweak vacuum symmetry breaking, and this will again put particle theory on the right track. If this does not happen, it is likely to be at least another decade if not longer before another chance for progress comes along, perhaps with the construction of a new linear electron–positron collider.

The science writer John Horgan stirred up considerable controversy in 1996 with the publication of his book *The End of Science.*[2] Horgan made a case for the idea that most of the big discoveries in science have been made, and that in danger of being reduced to just adding details to existing theories, scientists are more and more pursuing what he calls "ironic science." By ironic science Horgan means science pursued in a "speculative, post-empirical mode," something more like literary criticism that is inherently incapable of ever converging on the truth. In a more recent book, *Rational Mysticism,* Horgan applied another literary analogy to superstring theory, describing it as "little more than science fiction in mathematical form."[3] While in *The End of Science* he applied the idea of ironic science to developments in many different sciences, theoretical particle physics was his exhibit A, and he envisioned a future in which

> a few diehards dedicated to truth rather than practicality will practice physics in a nonempirical, ironic mode, plumbing the magical realm of superstrings and other esoterica and fretting about the meaning of quantum mechanics. The conferences of these ironic physicists, whose disputes cannot be experimentally resolved, will become more and more like those of that bastion of literary criticism, the Modern Language Association.[4]

These were considered fighting words by most physicists, and Horgan immediately became quite unpopular with them and with his employer, the magazine *Scientific American*. One reason for this was that he had very much struck a raw nerve, since to many physicists, there is already not much difference between a Modern Language Association (MLA) conference and one on superstring theory. Horgan's prediction was prescient since 1997, the year after his book appeared, was also the year of the first of what was to become an annual sequence of large international superstring theory conferences: "Strings 1997," in Amsterdam. This series of conferences has seen large numbers in attendance in recent years, with 445 participants in Strings 2002, at Cambridge University; 392 at Strings 2003, in Kyoto; 477 at Strings 2004, in Paris; and about 440 at Strings 2005, in Toronto. Unlike the MLA, these conferences are mostly held during the summer rather than the winter academic hiring season, so job interviews are not a big part of the scene. But just as at the MLA, not a single one of the string theory talks at any of these conferences involves any sort of experimentally verifiable prediction about the behavior of the physical world.

It seems to me that while Horgan put his finger accurately on what has been happening in particle theory as it has become a victim of its own success, the long-term future he envisions for the field is not a necessary one. The most important open problem of the subject, the nature of electroweak symmetry breaking, happens at an energy scale whose study should begin to be possible at the LHC. If the issue is not resolved then, there is still a good chance it would be at the next generation of particle accelerators. Various experiments concerning neutrino masses and mixing angles are in progress, and the data they provide over the next few years may be a new and important clue about how to go beyond the standard model.

More importantly, I am very familiar with the situation in a science that Horgan did not consider, the science of mathematics. Mathematics is not an experimental science and does not attempt predictions about the physical world, but it is a science nonetheless. It is a science that made great strides during the twentieth century, but nevertheless, there remains a great deal of mathematics that is not understood, and prospects are good that there is a lot more progress to come. The successes already seen over the past century have made the practice of the subject a much more difficult and technically sophisticated endeavor. In this sense, it is also a victim of its own success in much the same way as other sciences, but it is still far from the point where going any further would be beyond the reach of human intellect.

The past decade has seen the resolution of two of the most difficult longstanding problems known to mathematicians. The proof of Fermat's last theorem found by Andrew Wiles in 1994 resolved what was perhaps the best-known unsolved problem in mathematics, one that had resisted efforts of the best mathematicians for three centuries. More recently, in 2003, Grigori Perelman announced an outline of a proof of the most famous open problem in topology, the century-old Poincaré conjecture. In both cases the solution of these problems required Wiles and Perelman to devote seven or more years of their lives to the task, bringing to bear the full arsenal of modern mathematical techniques.

Traditionally, the two biggest sources of problems that motivate new mathematics have been the study of numbers and the study of theoretical physics. These function in some sense as experimental data for mathematicians, throwing out new puzzles that may perhaps be explained through new mathematical structures. We have seen the huge positive effect that quantum field theory has had on mathematics over the last twenty years, and this effect is likely to continue. Mathematics may some day be able to return the favor by giving physicists new mathematical techniques they can use to solve their problems, but I believe there are also other ways in which it can provide an important example to theoretical physicists.

Mathematicians have a very long history of experience with how to work in the speculative, postempirical mode that Horgan calls ironic science. What they learned long ago was that to get anywhere in the long term, the field has to insist strongly on absolute clarity of the formulation of ideas and the rigorous understanding of their implications. Modern mathematics may be justly accused of sometimes taking these standards too far, to the point of fetishizing them. Often, mathematical research suffers because the community is unwilling to let appear in print the vague speculative formulations that motivate some of the best new work, or the similarly vague and imprecise summaries of older work that are essential to any readable expository literature.

In 1993, after the wave of new ideas coming into mathematics from Witten's work was beginning to be absorbed into the subject, Arthur Jaffe and Frank Quinn published a cautionary article in the *Bulletin of the American Mathematical Society.*[5] They were worried that a great deal of "Speculative mathematics" was entering the literature, and pointed out some of the dangers involved in dealing with situations in which it was not clear what was rigorously proved and what wasn't. The article led to a vigorous debate, including the following comments from Sir Michael Atiyah:

> But if mathematics is to rejuvenate itself and break exciting new ground it will have to allow for the exploration of new ideas and techniques which, in their creative phase, are likely to be as dubious as in some of the great eras of the past. Perhaps we now have high standards of proof to aim at but, in the early stages of new developments, we must be prepared to act in more buccaneering style . . .
>
> What is unusual about the current interaction is that it involves front-line ideas both in theoretical physics and in geometry. This greatly increases its interest to both parties, but Jaffe–Quinn want to

emphasize the dangers. They point out that geometers are inexperienced in dealing with physicists and are perhaps being led astray. I think most geometers find this attitude a little patronizing: we feel we are perfectly capable of defending our virtue.[6]

To mathematicians, what is at issue here is how strongly to defend what they consider their central virtue, that of rigorously precise thought, while realizing that a more lax set of behaviors is at times needed to get anywhere. Physicists have traditionally never had the slightest interest in this virtue, feeling they had no need for it. This attitude was justified in the past when there were experimental data to keep them honest, but now perhaps there are important lessons they can learn from the mathematicians. To be really scientific, speculative work must be subject to a continual evaluation as to what its prospects are for getting to the point of making real predictions. In addition, every effort must be made to achieve precision of thought wherever possible and always to be clear about exactly what is understood, what is not, and where the roadblocks to further understanding lie.

The mathematics literature often suffers from being either almost unreadable or concerned ultimately with not very interesting problems, but it is hard to believe that a mathematics journal would publish anything that made as little sense as the five papers of the Bogdanovs that the referees of physics journals found acceptable. The Bogdanov affair provides strong evidence that the speculative parts of theoretical physics have become so infected with incoherent thought and argument that many of its practitioners have given up even trying to insist that things make sense. This is a deadly situation for a field that now primarily deals in speculative ideas and can no longer rely upon experimental results.

If the field of particle theory does somehow manage to make an honest evaluation of the superstring theory program, then its own internal procedures should automatically lead to an end to the reward and encouragement of failed ideas. Hiring committees and granting agencies will stop hiring and funding researchers who won't search for new things to work on, and who insist on following a failed program. If this does not happen and if such an honest evaluation continues to be evaded, it may be time for concerned outsiders such as other members of physics departments and responsible administrators at the DOE and NSF to take matters into their own hands. The power to change the direction of research in particle theory lies in the hands of a small number of faculty committees and a couple of government offices. Dramatic effects could be seen quickly should they choose to exercise this power to bring about change.

At the same time, the particle theory community should be asking itself how it got to this point, and what can be done to get out of the present situation. Perhaps some structural adjustments in how research is organized are called for. Some possibilities that should be looked at include lengthening the term of postdocs to allow young researchers time to get somewhere with new ideas, and graduate student "birth control" to make the job situation a less brutal one for young physicists.

Several years before Horgan's *The End of Science,* physicist David Lindley, in his 1993 book *The End of Physics: The Myth of a Unified Theory,*[7] warned that in its search for a unified theory, physics was in danger of becoming mythology rather than science. Lindley and many other physicists see superstring theory as based on pure mathematics, relying on aesthetic judgments to measure progress. They accept the claim of superstring theorists that the theory is a beautiful and elegant one, and criticize this reliance on mathematical beauty as somehow keeping theorists from connecting their ideas with anything experimentally observable. In this book I have tried to show that this is a misguided point of view, one that is based on not looking closely enough into why superstring theory has not been able to make any predictions.

The beauty and elegance of superstring theory lies in the hopes and dreams of its practitioners, hopes and dreams that are vanishing as every year it becomes more and more unlikely that they are ever to be realized. Superstring theorists would like to believe that someday a simple equation, beautiful physical idea, or fundamental symmetry principle will be found that will explain the intricate structures they have been studying. The present situation of the field is that no such thing is actually in sight despite more than twenty years of effort looking for it. Those who have eloquently described the elegance and beauty of superstring theory are quite right that these are characteristics that a successful fundamental physical theory almost surely will have, but they often fail to distinguish dreams and reality. The current best picture of the world provided by actual existing superstring theory is neither beautiful nor elegant. The ten- and eleven-dimensional supersymmetric theories actually used are very complicated to write down precisely. The six- or seven-dimensional compactifications of these theories necessary to try to make them look like the real world are both exceedingly complex and exceedingly ugly.

At a conference at Harvard entitled "The Unity of Mathematics" in September 2003, Atiyah gave a talk on "The Interaction Between Geometry and Physics." In recent years he has collaborated with Witten in work on M-theory and remains an admirer of Witten and string theory, but he hopes for better things to come:

> If we end up with a coherent and consistent unified theory of the universe, involving extremely complicated mathematics, do we believe that this represents "reality"? Do we believe that the laws of nature are laid down using the elaborate algebraic machinery that is now emerging in string theory? Or is it possible that nature's laws are much deeper, simple yet subtle, and that the mathematical description we use is simply the best we can do with the tools we have? In other words, perhaps we have not yet found the right language or framework to see the ultimate simplicity of nature.

One other prediction he offered was that the geometry of spinors would be one place to look for the new geometrical structures that are needed.

Atiyah is one of the greatest mathematicians of the second half of the twentieth century, and very much influenced by Weyl, one of the greatest of the first half. Their careers hardly overlapped at all, but in an interview Atiyah says,

The person I admire most is Hermann Weyl. I have found that in almost everything I have ever done in mathematics, Hermann Weyl was there first . . .

For many years whenever I got into a different topic I found out who was behind the scene, and sure enough, it was Hermann Weyl. I feel my center of gravity is in the same place as his. Hilbert was more algebraical; I don't think he had quite the same geometrical insights. Von Neumann was more analytical and worked more in applied areas. I think Hermann Weyl is clearly the person I identify with most in terms of mathematical philosophy and mathematical interests.[8]

Much of the argument of this book has been of a very negative nature, criticizing the superstring theory program as a failed and overhyped project. More positively, I have also tried to explain what I see as an important lesson that each generation of physicists since the advent of quantum mechanics seems to need to learn anew. This lesson is the importance of symmetry principles, expressed in the mathematical language of group representation theory. Quantum mechanics loses much of its mystery and becomes a very natural way of thinking when one works in this language. The underlying source of the problems of superstring theory is that the theory is not built on a fundamental symmetry principle or expressed within the language of representation theory. Unless some way can be found to rework the theory into a form in which this is possible, the lesson of history is that it is never going to lead anywhere.

Very late in his life, Weyl wrote a popular book entitled *Symmetry*,[9] which was as much about art and beauty as about mathematics. In it he made the case that the notion of symmetry was central to classical artistic notions of beauty, beginning with a discussion of the case of symmetry under reflection. For him, the mathematical idea of a representation of a group by symmetries was a precise embodiment of the ideas of elegance and beauty. If one takes Weyl's point of view seriously, to search for a more beautiful physical theory than the standard model, one must do one of two things. One must either find new symmetry groups beyond those already known, or one must find more powerful methods for exploiting the mathematics of representation theory to obtain physical understanding.

One of the great insights of the standard model is the importance of the group of gauge symmetries, and it is a remarkable fact that in four space-time dimensions virtually nothing is known about the representations of this infinite-dimensional group. The traditional reason that physicists don't think this is important is that they believe that only things that are invariant under gauge transformations matter, or in other words, that only the trivial representation is needed. Thinking about things this way may very well turn out to be as misguided as thinking that the vacuum state of a quantum field theory cannot be interesting. Similarly, the fundamental principle of general relativity is that of invariance under the group of general coordinate transformations (diffeomorphisms), and little is known about the representation theory of this group. Perhaps the true secret of quantum gravity can be found once the representation theory of these gauge and diffeomorphism groups is better understood.

These speculations about the possibility of using representation theory to go beyond the standard model may of course easily turn out to be completely wrongheaded. I am convinced, however, that any further progress toward understanding the most fundamental constituents of the universe will require physicists to abandon the now ossified ideology of supersymmetry and superstring theory that has dominated the last two decades. Once they do so, one thing they may discover is that the marvelously rich interaction of quantum field theory and mathematics that has already so revolutionized both subjects was just a beginning.

Acknowledgments

My knowledge about the topics discussed in this book is the product of many years spent in both the mathematics and physics communities, and I have benefited over this time from conversations on these topics with a large number of mathematicians and physicists. By now, this number is so large that I am afraid that I am able to remember and thank only those who have been of help during the last few years while I have been writing this book and seeing it through to publication.

Of important assistance have been string theorists and other string-friendly physicists who have been willing to share their expertise with me as part of discussions about string theory in various Internet venues. I have learned a great deal from these discussions, so even though many of these physicists are not likely to be very happy with this book, their efforts have made it a much better one than would otherwise have been possible. For help of this kind, I would like specifically to thank Aaron Bergman, Sean Carroll, Jacques Distler, Robert Helling, Clifford V. Johnson, Lubos Motl, Moshe Rozali, and Urs Schreiber.

During the past few years I have been grateful to the many individuals who have contacted me, sometimes anonymously, with expressions of support and enthusiasm, often with an interesting story of one sort or another. Finding out that my concerns about the present state of particle theory are so widely shared has been an important incentive for the writing of this book.

I would also like to thank Roger Astley, of Cambridge University Press, together with Jim Lepowsky and several anonymous referees. While publication by Cambridge ultimately didn't work out, going through the process there ultimately turned the original manuscript into a much better book. At a later stage, Binky Urban provided excellent advice about what should be done to make the book a commercial success, much of which I fear I haven't taken.

Professor Karl von Meyenn provided me with the reference indicating that the attribution of the phrase "not even wrong" to Pauli is not as apocryphal as I had feared, together with useful comments about the context of the phrase and the influence of the Vienna Circle on Pauli's thinking.

John Horgan has provided encouragement and advice, together with interesting discussion of his views on some of the topics treated here.

Among friends and supporters who have listened to me go on far too long about string theory and some of the other subjects treated here, I would like especially to thank Oisin McGuinness, Nathan Myhrvold, Peter Orland, and Eric Weinstein, as well as my Columbia colleagues Bob Friedman, John Morgan, D. H. Phong, and Michael Thaddeus, whose interest and support have been invaluable.

It has been an honor to have the opportunity to discuss in person or via e-mail some of the issues treated here with some truly great mathematicians and physicists, all of whom have provided helpful advice. These include Gerard 't Hooft, Lee Smolin, and Martin Veltman, as well as two people whose work has had an overwhelmingly large influence on my understanding of mathematics and physics: Sir Michael Atiyah and Edward Witten. While many of them may very strongly disagree with the point of view from which I am writing, whatever value there is in it owes a great deal to them.

Another great physicist, Sir Roger Penrose, was of critical help in encouraging this project and helping to find it a publisher, for which I am exceedingly grateful.

My editors, Will Sulkin and Richard Lawrence at Jonathan Cape, and Bill Frucht and David Kramer at Basic Books, have been a pleasure to work with, and I hope they won't regret their willingness to help me write and publish exactly the book I have wanted to write, one that may be more intellectually uncompromising than most publishers would find comfortable.

Finally, I owe an immeasurable debt to Ellen Handy. Her extensive editorial assistance over many years has created a text of far higher quality than would otherwise have been conceivable. More importantly, her love, selfless encouragement, and unwavering belief in me have made possible the writing of this book, and much else besides.

NOTES

PREFACE

1. R. Peierls. *Biographical Memoirs of Fellows of the Royal Society* 5 (February 1960), p. 186.

2. W. Heisenberg. *Across the Frontiers*. Harper and Row, 1974.

3. H. Weyl. *Gruppentheorie und Quantenmechanik*. S. Hirzel, 1928.

Chapter 1

1. A. Einstein. Autobiographical Notes, in *Albert Einstein: Philosopher–Scientist*. P. Schilpp, ed. Open Court Publishing, 1969, p. 63.

2. L. Susskind. http://www.edge.org/3rd_culture/susskind03/susskind_index.html.

3. L. Susskind. *The Cosmic Landscape: String Theory and the Illusion of Intelligent Design*. Little, Brown and Company, 2005.

4. D. Gross, October 11, 2003. http://www.phys.cwru.edu/events/cerca_video_archive.php.

Chapter 2

1. Lawrence and the Cyclotron. Online exhibit. Center for History of Physics, http://www.aip.org/history/lawrence.

2. J. Cramer. The Decline and Fall of the SSC. *Analog Science Fiction and Fact*. May 1997.

3. H. Wouk. *A Hole in Texas*. Little, Brown and Company, 2004.

4. F. Close, S. Sutton, and M. Marten. *The Particle Explosion*. Oxford University Press, 1987.

5. S. Weinberg. *Discovery of Subatomic Particles*. W. H. Freeman, 1983.

6. P. Galison. *Image and Logic*. University of Chicago Press, 1997.

7. S. Traweek. *Beamtimes and Lifetimes: The World of High Energy Physicists*. Harvard University Press, 1990.

Chapter 3

1. R. Crease and C. Mann. *The Second Creation*. Macmillan, 1986, p. 52.

2. P. A. M. Dirac. *Wisconsin State Journal*, April 31, 1929. Quoted in Schweber, *QED and the Men Who Made It*. Princeton University Press, 1994. pp. 19–20.

3. W. Moore. *A Life of Erwin Schrödinger*. Cambridge University Press, 1994, p. 111.

4. E. Schrödinger. Quantisierung als Eigenwert Problem: Erste Mitteilung. *Annalen der Physik* 79 (1926), p. 361.

5. W. Moore. *Life of Erwin Schrödinger*. Cambridge University Press, 1994, p. 138.

6. W. Moore. *A Life of Erwin Schrödinger*. p. 128.

7. H. Weyl. Emmy Noether, *Scripta Mathematica* 3 (1935), pp. 201–220.

8. H. Weyl. *Gruppentheorie und Quantenmechanik* S. Hirzel, Leipzig, 1928.

9. E. Wigner. *The Recollections of Eugene P. Wigner*. Plenum, 1992, p. 118.

10. C. N. Yang. Hermann Weyl's Contributions in Physics. In K. Chandrasekharan, ed., *Hermann Weyl: 1885–1985*. Springer-Verlag, 1986.

11. P. A. M. Dirac. *The Principles of Quantum Mechanics*. Oxford University Press, 1958.

12. T. Hey and P. Walters. *The New Quantum Universe*. Cambridge University Press, 2003.

13. V. Guillemin. *The Story of Quantum Mechanics*. Charles Scribner's Sons, 1968.

14. J. Mehra and H. Rechenberg. *The Historical Development of Quantum Theory* (six volumes). Springer-Verlag, 1982–2001.

15. A. Zee. *Fearful Symmetry*. Macmillan, 1986.

16. F. Wilczek and B. Devine. *Longing for the Harmonies*. W. W. Norton, 1989.

17. M. Livio. *The Equation That Couldn't Be Solved*. Simon and Schuster, 2005.

18. E. Wigner. *Symmetries and Reflections*. MIT Press, 1970.

19. H. Weyl. *Symmetry*. Princeton University Press, 1952.

20. B. Hall. *Lie Groups, Lie Algebras, and Representations*. Springer-Verlag, 2003.

21. B. Simon. *Representations of Finite and Compact Groups*. American Mathematical Society, 1996.

22. W. Rossman. *Lie Groups*. Oxford University Press, 2002.

23. T. Hawkins. *Emergence of the Theory of Lie Groups*. Springer-Verlag, 2000.

24. R. Penrose. *The Road to Reality: A Complete Guide to the Laws of the Universe*. Jonathan Cape, 2004.

Chapter 4

1. A. Watson. *The Quantum Quark*. Cambridge University Press, 2004, p. 325.

2. B. Lautrup and H. Zinkernagel. *Studies in History and Philosophy of Modern Physics* 30 (1999) 85–119.

3. T. Kinoshita. New Value of the α^3 Electron Anomalous Magnetic Moment. *Phys. Rev. Lett.* 75 (1995) 4728–4731.

4. R. Jost. Quoted in R. Streater and A. Wightman. *PCT, Spin and Statistics, and All That*. Benjamin, 1964, p. 31.

5. S. Schweber. *QED and the Men Who Made It*. Princeton University Press, 1994.

6. S. Schweber. *QED and the Men Who Made It*. p. 491.

7. J. Bjorken and S. Drell. *Relativistic Quantum Mechanics*. McGraw-Hill, 1964.

8. J. Bjorken and S. Drell. *Relativistic Quantum Fields*. McGraw-Hill, 1965.

9. C. Itzykson and J.-B. Zuber. *Quantum Field Theory*. McGraw-Hill, 1980.

10. P. Ramond. *Field Theory: A Modern Primer*. Benjamin/Cummings, 1981.

11. M. Peskin and D. Schroeder. *An Introduction to Quantum Field Theory*. Westview Press, 1995.

12. A. Zee. *Quantum Field Theory in a Nutshell*. Princeton University Press, 2003.

13. R. Feynman. *QED: The Strange Theory of Light and Matter*. Princeton University Press, 1986.

14. T. Y. Cao, ed. *Conceptual Foundations of Quantum Field Theory*. Cambridge University Press, 1999.

15. P. Teller. *An Interpretive Introduction to Quantum Field Theory*. Princeton University Press, 1995.

16. S. Weinberg. *The Quantum Theory of Fields, I, II, and III*. Cambridge University Press, 1995, 1996, 2000.

Chapter 5

1. H. Weyl. Gravitation and Electricity. *Sitzungsber. Preuss. Akad. Berlin* (1918), p. 465.

2. H. Weyl. *Space–Time–Matter*. Dover, 1952.

3. V. Raman and P. Forman. *Hist. Studies Phys. Sci.* 1 (1969), p. 291.

4. Translation from C. N. Yang. Square Root of Minus One, Complex Phases and Erwin Schrödinger. In C. W. Kilmister, ed. *Schrödinger: Centenary Celebration of a Polymath*. Cambridge University Press, 1987.

5. H. Weyl. *Zeit. f. Phys.* 56 (1929), p. 330.

6. R. Bott. On Some Recent Interactions between Mathematics and Physics. *Canad. Math. Bull.* 28 (1985), pp. 129–164.

7. L. O'Raifeartaigh. *The Dawning of Gauge Theory*. Princeton University Press, 1997.

Chapter 6

1. R. Crease and C. Mann. *The Second Creation*. Macmillan, 1986.

2. M. Riordan. *The Hunting of the Quark*. Simon and Schuster, 1987.

3. G. Johnson. *Strange Beauty: Murray Gell-Mann and the Revolution in 20th-Century Physics*. Vintage, 2000.

4. H. Georgi. *Lie Algebras in Particle Physics*. Benjamin/Cummings, 1982, p. 155.

5. H. Georgi. *Lie Algebras in Particle Physics*. p. xxi.

6. D. Gross. *Physics Today,* December 2004, p. 22.

7. M. Riordan. *The Hunting of the Quark.* Simon and Schuster, 1987.

8. R. Crease and C. Mann. *The Second Creation.* Macmillan, 1986.

9. L. M. Brown, M. Dresden, and L. Hoddeson. *Pions to Quarks: Particle Physics in the 1950s: Based on a Fermilab Symposium.* Cambridge University Press, 1989.

10. L. Hoddeson, L. Brown, M. Riordan, and M. Dresden. *The Rise of the Standard Model: Particle Physics in the 1960s and 1970s.* Cambridge University Press, 1997.

11. S. Weinberg. The Making of the Standard Model in G. 't Hooft. *50 Years of Yang–Mills Theory.* World Scientific, 2005.

12. G. 't Hooft. *50 Years of Yang–Mills Theory.* World Scientific, 2005.

Chapter 7

1. M. Riordan. *The Hunting of the Quark.* Simon and Schuster, 1987.

2. G. Taubes. *Nobel Dreams: Power, Deceit, and the Ultimate Experiment.* Random House, 1987.

3. G. 't Hooft. *In Search of the Ultimate Building Blocks.* Cambridge University Press, 1996.

4. M. Veltman. *Facts and Mysteries in Elementary Particle Physics.* World Scientific, 2003.

5. D. Gross, D. Politzer, and F. Wilczek. http://nobelprize.org/physics/laureates/ 2004.

Chapter 9

1. G. Kane. *Supersymmetry: Unveiling the Ultimate Laws of Nature.* Perseus, 2000.

2. S. Hawking. Is the End in Sight for Theoretical Physics? In *Black Holes and Baby Universes and Other Essays.* Bantam, 1994.

Chapter 10

1. Curtis Callan, Jr. Princeton, February 27, 2003.

2. M. Atiyah. *Michael Atiyah Collected Works.* Volume 5: *Gauge Theories.* Oxford University Press, 1988.

3. E. Witten. Michael Atiyah and the Physics/Geometry Interface. *Asian J. Math.* 3 (1999), pp. lxi–lxiv.

4. R. Bott. Morse Theory Indomitable, *Publ. Math. IHES* 68 (1988).

5. M. Atiyah. New Invariants of 3 and 4 Dimensional Manifolds. In *The Mathematical Heritage of Hermann Weyl.* American Mathematical Society, 1989.

6. E. Witten. Michael Atiyah and the Physics/Geometry Interface.

7. A. Jaffe. The Role of Rigorous Proof in Modern Mathematical Thinking. In *New Trends in the History and Philosophy of Mathematics.* T. H. Kjeldsen, S. A. Pedersen, and I. M. Sonne-Handsen, eds. Odense University Press, 2004, pp. 105–116.

8. M. Atiyah. *Geometry of Yang–Mills Fields.* Scuola Normale Superiore, Pisa, 1979.

9. S. Coleman. *Aspects of Symmetry: Selected Erice Lectures.* Cambridge University Press, 1985.

10. H. Rothe. *Lattice Gauge Theories: An Introduction.* World Scientific, 1997.

11. S. Treiman, R. Jackiw, B. Zumino, and E. Witten. *Current Algebra and Anomalies.* Princeton University Press, 1985.

12. M. Atiyah. Anomalies and Index Theory. In *Lecture Notes in Physics* 208. Springer-Verlag 1984.

13. M. Atiyah. Topological Aspects of Anomalies. In *Symposium on Anomalies, Geometry and Topology.* World Scientific, 1984.

14. S. Coleman. *Aspects of Symmetry: Selected Erice Lectures.* Cambridge University Press, 1985.

15. P. Deligne et al. *Quantum Fields and Strings: A Course for Mathematicians* (two volumes). American Mathematical Society 1999.

16. K. Hori et al. *Mirror Symmetry.* American Mathematical Society, 2003.

Chapter 11

1. B. Greene. *The Elegant Universe.* W. W. Norton, 1999.

2. B. Greene. *The Fabric of the Cosmos.* Knopf, 2004.

3. M. Kaku. *Hyperspace.* Anchor, 1995.

4. M. Kaku. *Beyond Einstein: The Cosmic Quest for the Theory of the Universe.* Anchor, 1995.

5. M. Kaku. *Parallel Worlds.* Doubleday, 2004.

6. W. Pauli. Difficulties of Field Theories and of Field Quantization. In *Report of an International Conference on Fundamental Particle and Low Temperatures.* The Physical Society, 1947.

7. G. Chew. *The Analytic S Matrix: A Basis for Nuclear Democracy.* Benjamin, 1966, p. 95.

8. D. Gross. Asymptotic Freedom, Confinement and QCD. In *History of Original Ideas and Basic Discoveries in Particle Physics.* H. Newman and T. Ypsilantis, eds. Plenum Press, 1996.

9. A. Pickering. From Field Theory to Phenomenology: The History of Dispersion Relations. In *Pions to Quarks: Particle Physics in the 1950s.* L. Brown, M. Dresden, and L. Hoddeson, eds. Cambridge, 1989.

10. D. Gross. Asymptotic Freedom, Confinement and QCD.

11. F. Capra. *The Tao of Physics.* Shambhala, 1975.

12. F. Capra. *The Tao of Physics.* 3rd edition. Shambhala, 1991, p. 257.

13. F. Capra. *The Tao of Physics,* p. 9.

14. F. Capra. *The Tao of Physics,* p. 315.

15. F. Capra. *The Tao of Physics,* p. 327.

16. L. Susskind. In L. Hoddeson, L. Brown, M. Riordan, and M. Dresden, eds. *The Rise of the Standard Model: Particle Physics in the 1960s and 1970s.* Cambridge University Press, 1997, p. 235.

17. J. Schwarz. In *History of Original Ideas and Basic Discoveries in Particle Physics*. H. Newman and T. Ypsilantis, eds. Plenum Press, 1996, p. 698.

18. E. Witten. $D = 10$ Superstring Theory. In H. A. Weldon, P. Langacker, and P. J. Steinhardt, eds. *Fourth Workshop on Grand Unification*. Birkhäuser, 1983, p. 395.

19. J. Frank Adams. Finite H-Spaces and Lie Groups. *J. Pure and Applied Algebra* 19 (1980), pp. 1–8.

20. M. Reid. Update on 3-Folds. *Proceedings of the ICM, Beijing 2002*. Vol. 2, pp. 513–524.

21. E. Witten. Magic, Mystery, and Matrix. *Notices of the AMS* 45 (1998), pp. 1124–1129.

22. S. Glashow. Interview in *NOVA* special *The Elegant Universe*. Public Broadcasting Service, 2003.

23. G. Kane and M. Shifman, eds. *The Supersymmetric World: The Beginnings of the Theory*. World Scientific, 2001.

24. J. Schwarz. In *History of Original Ideas and Basic Discoveries in Particle Physics*. p. 698.

25. M. Green, J. Schwarz, and E. Witten. *Superstring Theory* (two volumes). Cambridge University Press, 1988.

26. J. Polchinski. *String Theory* (two volumes). Cambridge University Press, 1998.

27. B. Zwiebach. *A First Course in String Theory*. Cambridge University Press, 2004.

28. C. Johnson. *D-Branes*. Cambridge University Press, 2003.

29. L. Randall. *Warped Passages: Unraveling the Mysteries of the Universe's Hidden Dimensions*. Ecco, 2005.

Chapter 12

1. G. Kane and M. Shifman. *The Supersymmetric World*. World Scientific, 2000.

2. SLAC SPIRES database.

3. S. Coleman. *Aspects of Symmetry: Selected Erice Lectures*. Cambridge University Press, 1985.

4. G. Kane. TASI Lectures: Weak Scale Supersymmetry: A Top-Motivated–Bottom-Up Approach. arXiv:hep-th/981210.

5. P. C. W. Davies and J. R. Brown. *Superstrings: A Theory of Everything?* Cambridge University Press, 1988.

6. L. Krauss. Talk at Asimov Panel Discussion, Museum of Natural History, New York, February 2001.

7. S. Glashow and B. Bova. *Interactions: A Journey Through the Mind of a Particle Physicist*. Warner Books, 1988, p. 25.

8. E. Witten. $D = 10$ Superstring Theory. In H. A. Weldon, P. Langacker, and P. J. Steinhardt, eds. *Fourth Workshop on Grand Unification*. Birkhäuser, 1983, p. 395.

9. T. Banks. Matrix Theory, arXiv:hep-th/9710231.

10. G. 't Hooft. *In Search of the Ultimate Building Blocks.* Cambridge University Press, 1997, p. 163.

11. L. Smolin. How Far Are We from the Quantum Theory of Gravity? arXiv:hep-th/9812104.

12. L. Smolin. *Three Roads to Quantum Gravity.* Basic Books, 2001.

13. M. Shifman. From Heisenberg to Supersymmetry. *Fortschr. Phys.* 50 (2002), pp. 5–7.

14. W. Heisenberg. *Introduction to the Unified Field Theory of Elementary Particles.* Interscience, 1966.

15. L. Krauss. *Hiding in the Mirror.* Viking, 2005.

16. D. Friedan. A Tentative Theory of Large Distance Physics. arXiv:hep-th/0204131.

17. M. Gell-Mann. Closing Talk at the Second Nobel Symposium on Particle Physics, 1986. *Physica Scripta* T15 (1987), p. 202.

18. J. Magueijo. *Faster Than the Speed of Light.* Perseus, 2003, p. 239–240.

19. J. Magueijo. *Faster Than the Speed of Light.* pp. 236–237.

Chapter 13

1. Voltaire. *Candide.* 1759.

2. T. Southern and M. Hoffenberg. *Candy.* Putnam, 1964.

3. E. Wigner. The Unreasonable Effectiveness of Mathematics in the Natural Sciences. *Comm. Pure Applied Mathematics* 13 (1960), pp. 1–14.

4. E. Baum. *What Is Thought?* MIT Press, 2004.

5. P. A. M. Dirac. The Evolution of the Physicist's Picture of Nature. *Scientific American,* May 1963, pp. 45–53.

6. J. Schwarz. Strings and the Advent of Supersymmetry: The View from Pasadena. In G. Kane and M. Shifman. *The Supersymmetric World.* World Scientific, 2000, pp. 16–17.

7. L. Susskind. *The Cosmic Landscape: String Theory and the Illusion of Intelligent Design.* Little, Brown and Company, 2005, p. 125.

8. L. Susskind. *The Cosmic Landscape: String Theory and the Illusion of Intelligent Design.* p. 377.

9. L. Susskind. *The Cosmic Landscape: String Theory and the Illusion of Intelligent Design.* p. 124.

10. A. Cho. String Theory Gets Real—Sort Of, *Science* 306, (November 26, 2004), pp. 1460–1462.

11. P. Anderson. www.edge.org/q2005/q05_10.html#andersonp.

Chapter 14

1. J. Polchinski. Talk given at the 26th SLAC Summer Institute on Particle Physics. arXiv:hep-th/9812104.

2. S. Hawking. *A Brief History of Time.* Bantam, 1988, p. 1.

3. C. Geertz. *The Interpretation of Cultures.* Basic Books, 1973, p. 28.

4. S. Glashow. Does Elementary Particle Physics Have a Future? In *The Lesson of Quantum Theory,* J. de Boer, E. Dal, and O. Ulfbeck, eds. Elsevier, 1986, pp. 143–153.

5. M. Kaku. Interview on the *Leonard Lopate Show,* WNYC, January 2, 2004.

Chapter 16

1. P. C. W. Davies and J. R. Brown. *Superstrings: A Theory of Everything?* Cambridge University Press, 1988, p. 148.

2. G. Taubes. *Nobel Dreams: Power, Deceit and the Ultimate Experiment.* Random House, 1986, pp. 254–255.

3. D. Gross. Talk at AAAS session on "The Coming Revolutions in Particle Physics." February 16, 2001, http://www.aaas.org/meetings/2001/6128.00.htm.

4. J. Polchinski. Talk given at the 26th SLAC Summer Institute on Particle Physics. arXiv:hep-th/9812104.

5. A. Jha. String Fellows. *The Guardian,* January 20, 2005.

6. J. Glanz. Even without Evidence, String Theory Gains Influence. *New York Times,* March 13, 2001.

7. B. Greene. *The Elegant Universe.* W. W. Norton, 1999.

8. B. Greene. *The Fabric of the Cosmos.* Knopf, 2004.

9. G. Johnson. Physicists Finally Find a Way to Test Superstring Theory. *New York Times,* April 4, 2002.

10. B. Zwiebach. *A First Course in String Theory.* Cambridge University Press, 2004.

11. S. Gruner, J. Langer, P. Nelson, and V. Vogel. What Future Will We Choose for Physics? *Physics Today.* December 1995, pp. 25–30.

12. M. Henly and R. Chu. AIP Society Membership Survey: 2000.

13. Particle Data Group. *The 2002 Census of U.S. Particle Physics.*

14. P. Oddone and D. Vaughan. *Survey of High-Energy Physics Support at U.S. Universities.* DOE, 1997.

15. Theoretical Particle Physics Jobs Rumor Mill. http://www.physics.wm.edu/~calvin.

16. M. Raussen and C. Skau. Interview with Michael Atiyah and Isadore Singer. *Notices of the AMS* 52 (2005), pp. 228–231.

17. M. Duff and S. Blackburn. Looking for a Slice of the American Pie. *Times Higher Education Supplement,* July 30, 1999.

Chapter 17

1. S. Kachru, R. Kallosh, A. Linde, and S. Trivedi. De Sitter Vacua in String Theory. arXiv:hep-th/0301240.

2. L. Susskind. Talk at conference in honor of Albert Schwarz, University of California at Davis, May 15, 2004.

3. S. Weinberg. *Phys. Rev. Lett.* 59 (1987), p. 2607.

4. L. Susskind. http://www.edge.org/3rd_culture/susskind03/susskind_index.html.

5. L. Susskind. Stanford University Press Release, February 15, 2005.

6. L. Susskind. *The Cosmic Landscape: String Theory and the Illusion of Intelligent Design.* Little, Brown and Company, 2005.

7. D. Overbye. One Cosmic Question, Too Many Answers. *New York Times,* September 2, 2003.

8. E. Witten. Talk at KITP, Santa Barbara, October 9, 2004.

9. D. Gross. Where Do We Stand in Fundamental Theory? Abstract for talk at Nobel Symposium, August 2003.

10. J. Polchinski. sci.physics.strings posting, April 7, 2004.

11. W. Lerche, sci.physics.strings posting, April 6, 2004.

12. M. Douglas. Statistical Analysis of the Supersymmetry Breaking Scale. arXiv:hep-th/0405279.

13. L. Susskind. Naturalness and the Landscape. arXiv:hep-ph/0406197.

14. A. M. Polyakov. Confinement and Liberation. arXiv:hep-th/0407209.

15. L. Susskind. *The Cosmic Landscape: String Theory and the Illusion of Intelligent Design.* pp. 192–193.

16. P. Steinhardt. http://www.edge.org/q2005/q05_print.html#steinhardt.

17. D. Overbye. One Cosmic Question, Too Many Answers. *New York Times,* September 2, 2003.

18. A. Cho. "String Theory Gets Real—Sort of," *Science* 306, (November 26, 2004) pp. 1460–1462.

Chapter 18

1. S. Weinberg. *The First Three Minutes: A Modern View of the Origin of the Universe.* Basic Books, 1977.

2. L. Smolin. *Three Roads to Quantum Gravity.* Basic Books, 2001.

3. C. Rovelli. *Quantum Gravity.* Cambridge University Press, 2004.

4. R. Penrose. *The Road to Reality: A Complete Guide to the Laws of the Universe.* Jonathan Cape, 2004.

5. A. Connes. *Noncommutative Geometry.* Academic Press, 1994.

Chapter 19

1. D. Friedan. A Tentative Theory of Large Distance Physics. arXiv:hep-th/0204131.

2. J. Horgan. *The End of Science: Facing the Limits of Knowledge in the Twilight of the Scientific Age.* Addison-Wesley, 1996.

3. J. Horgan. *Rational Mysticism.* Houghton-Mifflin, 2003, p. 175.

4. J. Horgan. *The End of Science: Facing the Limits of Knowledge in the Twilight of the Scientific Age.* p. 91.

5. A. Jaffe and F. Quinn. Theoretical Mathematics: Toward a Cultural Synthesis of Mathematics and Theoretical Physics, *Bull. A.M.S., 29* (1993), pp. 1–13.

6. M. Atiyah. Response to "Theoretical Mathematics . . . " *Bull. A.M.S.* 30 (1994), pp. 178–179.

7. D. Lindley D. *The End of Physics.* Basic Books, 1993.

8. M. Atiyah. An Interview with Michael Atiyah. *Mathematical Intelligencer 6* (1984), pp. 9–19.

9. H. Weyl. *Symmetry.* Princeton University Press, 1952.

INDEX